T0291356

The Business of Humanity makes the business case for why a more human-centered capitalism is important as a long-term business strategy and offers a formula for sustainable economic behavior that could benefit billions of people. Every CEO and politician should read this book.

Grant Oliphant
President, The Heinz Endowments

With specific examples, the authors challenge businesses of all sizes and locations to question whether their focus is on the drivers of real customer value, progressive management concepts, and out-of-the-box thinking of what really drives long-term success.

Donald R. Beall
Chairman Emeritus, Rockwell
Fortune Magazine's "Star" of the Aerospace and Defense Industries

African businesses are compelled by a variety of factors to incorporate corporate social responsibility and strong humane management into their business models. This book lays out the inductive logic and empirical basis for following this business model for enhanced sustainable economic performance.

Prof. Dr. Bart O. Nnaji
Chairman and CEO of Geometric Power Ltd.; former Minister of Power
and former Federal Minister of Science and Technology in Nigeria

The Business of Humanity demonstrates that profitable competitive business models are often enhanced when leaders consider the impact of their actions on broader humanity. It's a must read.

Thomas D. Hull III
Chief Financial Officer, Kewaunee Scientific Corporation

These scholars have developed a disruptive innovation model that can transform the business world. This book is the opportunity for businessmen and policy makers to implement this revolution all over the world.

Roberto Zarama
Professor, Universidad de los Andes

Organizations seeking to grow in new or emerging markets would benefit from applying the principles in this book.

Joel Ross

CEO, Universal Electric Corporation

Companies across the world, for a variety of reasons, are committing to incorporating social responsibility into their business models and finding that their profits are growing and their long-term sustainability is enhanced—building "humanity" into their business models as the driver of economic, environmental, and social sustainability. This fascinating development is a widely observable global phenomenon.

The "Business of Humanity®" (BoH) Proposition is the synthesis of counterintuitive but simple and powerful ideas about how companies can add value in today's globalized and fast-changing world. The task of BoH Strategies is to overcome three critical challenges characterizing today's business environment, namely disruptive technologies, conflicted stakeholders, and unknowable futures. BoH Strategies are designed to convert these challenges into opportunities for enhanced sustainability on all three dimensions—economic, environmental, and social. Written by leading experts with decades of experience, this book

- Provides a hands-on understanding of how to implement this powerful and rewarding approach to simultaneously add economic value and enhance social benefit
- Includes the experiences and approaches of highly regarded business executives and successful organizations
- Responds to the critical challenges created by three environmental mega forces—the inevitability of globalization, the imperative of innovation, and the importance of shared value

This book is based on lessons drawn from the real world and provides a compelling rationale for the power of the BoH Proposition. The pragmatic framework and process offered enable companies to develop and confidently implement value-adding strategies based on the BoH Proposition.

The Business of Humanity

Strategic Management
in the Era of Globalization,
Innovation, and Shared Value

The Business of Humanity

Strategic Management in the Era of Globalization, Innovation, and Shared Value

John C. Camillus
Bopaya Bidanda
N. Chandra Mohan

Routledge
Taylor & Francis Group

Routledge
Taylor & Francis Group
711 Third Avenue
New York, NY 10017

Routledge
Taylor & Francis Group
27 Church Road
Hove, East Sussex BN3, 2FA

© 2017 by John C. Camillus and Bopaya Bidanda
Routledge is an imprint of Taylor & Francis Group, an Informa business

International Standard Book Number-13: 978-1-138-19746-6 (Hardback)

For permission to photocopy or use material electronically from this work, please access www
.copyright.com (http://www.copyright.com/) or contact the Copyright Clearance Center, Inc.
(CCC), 222 Rosewood Drive, Danvers, MA 01923, 978-750-8400. CCC is a not-for-profit organization that provides licenses and registration for a variety of users. For organizations that have
been granted a photocopy license by the CCC, a separate system of payment has been arranged.

Trademark Notice: Product or corporate names may be trademarks or registered trademarks,
and are used only for identification and explanation without intent to infringe.

Library of Congress Cataloging-in-Publication Data

Library of Congress Cataloging-in-Publication Data
Names: Camillus, John C., author. | Bidanda, Bopaya, author. | Mohan, N. Chandra,
author.
Title: The business of humanity : strategic management in the era of globalization,
innovation, and shared value / John C. Camillus, Bopaya Bidanda, N. Chandra Mohan.
Description: Boca Raton, FL : CRC Press, 2018.
Identifiers: LCCN 2017002138 | ISBN 9781138197466 (hardback : alk. paper)
Subjects: LCSH: Social responsibility of business. | Strategic planning. | Globalization--
Economic aspects. | Technological innovations--Economic aspects. | Values.
Classification: LCC HD60 .C324 2018 | DDC 658.4/012--dc23
LC record available at https://lccn.loc.gov/2017002138

Visit the Taylor & Francis Web site at
http://www.taylorandfrancis.com

and the Routledge Web site at
http://www.routledgementalhealth.com

Printed and bound in Great Britain by
TJ International Ltd, Padstow, Cornwall

Contents

List of Figures

List of Tables

Preface

There is a growing global phenomenon that is a fascinating development to observe and study. Companies across the world, for a variety of reasons, are committing to incorporating social responsibility, responding to social concerns and issues as an integral and guiding component of their business models, and finding, as a result, that their profits are growing and their long-term sustainability is enhanced. We believe that the business world is approaching a tipping point—building "humanity" into business models as a driver of economic, environmental, and social sustainability.

The fundamental proposition of this book—the Business of Humanity (BoH) Proposition—based on inductive logic and empirical evidence, is that:

> Strategic decision-making that employs criteria falling under the rubric of "humanity," in its two dimensions of "humaneness" and "humankind," leads to superior economic performance and sustainability.

The BoH Proposition leads to *humane* decisions and actions that, we argue, will increase the profits generated and, importantly, reduce missteps when entering a new and untested market. The BoH Proposition motivates companies to respond to *humankind* in business decisions by recognizing and tapping into the innovation and revenue growth potential of emerging markets at the bottom of the pyramid.

There is growing evidence that this is an idea whose time has come. We have observed an increasing number of companies committing to build "humanity" into their business models as a driver for economic, environmental, and social sustainability. The theoretical basis, illuminating examples, and an operational framework for incorporating "humanity" in business decision-making are the foci of our book.

A decade ago, we founded the Business of Humanity® Project at the Katz Graduate School of Business and the Swanson School of Engineering at the University of Pittsburgh. We serve as the principal investigators of this Project, which is permanently funded by the Beall Family Foundation and has received significant support from other foundations, universities, corporations, and government agencies.

The BoH Project has been documenting and studying how "humanity" in strategic decision-making can enhance profits and sustainability. In this endeavor, we have engaged academics and practitioners in many countries. Analytical tools and management frameworks have been developed through empirical research that has spanned the globe. These tools and frameworks have been refined and tested through international conferences engaging both practitioners and academics, and through graduate courses offered in North and South America, Europe, and Asia.

A parallel research stream in which we have been engaged, which has significantly informed the content of this book, addressed the challenge of wicked problems. Wicked problems defy ready description. They are viewed differently by diverse stakeholders. These are problems where no single optimal answer can be arrived at, where attempted solutions change the nature of the problem, and where unexpected developments will inevitably be encountered. We have developed and written about "Wicked Strategies®" that are designed to tame and indeed exploit such wicked problems to gain competitive advantage. Wicked Strategies work powerfully *in response* to wicked problems. The BoH Strategies that we offer in this book go further as they *preempt and indeed embrace* the circumstances that create wicked problems.

This book incorporates the lessons drawn from case studies around the world and our extensive—over a hundred companies on four continents—consulting experience. We offer a commonsense rationale for the power of the BoH Proposition. More importantly, we offer a pragmatic and simple framework and a structured process to enable companies to develop and confidently implement value-adding strategies based on the BoH Proposition.

While the incorporation of "humanity" into an organization's business model may improve its ethical stance and enhance its contribution to the common good, the BoH is not just about corporate social responsibility for its own sake. "Doing well by doing good" may appear to be a fuzzy and Pollyannaish idea, but, as we have elaborated in our book, the BoH approach is based on an informed understanding derived from the experience of businesses, and seeks to offer a reasoned response to the limitations of accounting numbers and traditional analytical techniques in meeting today's strategic challenges.

This book is intended to provide readers with a hands-on understanding of how to implement this powerful and rewarding approach to simultaneously adding economic value and enhancing social benefit. With its focus on the practical development and implementation of the techniques, systems, and structures that support BoH Strategies, the book is intended for practicing managers, consultants, and educators. For the practicing manager, it is intended to serve as a hands-on guide. For consultants, it will enable them to provide their clients with a powerful and proven approach to enhanced and sustained profitability. For educators, it will serve to prepare their students to better understand and effectively deal with the current and emerging challenges facing global business.

John C. Camillus
Bopaya Bidanda
N. Chandra Mohan
Pittsburgh and New Delhi

1

The Business of Humanity Proposition

There is a growing global phenomenon that has been, for many of us, at the periphery of our consciousness and, for some of us, has been a fascinating development to observe and study. Companies across the world, for a variety of reasons, are committing to incorporating social responsibility into their business models and finding that their profits are growing and their long-term sustainability is enhanced. We believe that the business world is at a tipping point because companies like Dow Chemical, Costco, and Ford in the Americas; Arvind, the Tata Group, Hitachi, and CavinKare in Asia; and Unilever and EMCO in Europe are explicitly and formally building "humanity" into their business models as the driver of economic, environmental, and social sustainability.

Over the last several years, we have observed and closely studied many examples of companies that have paid explicit attention to "humanity." We believe we now have a sound, working understanding of how "humanity" can enhance profits and sustainability. This book shares what we have learned over the last several years, and is intended to provide you with a hands-on understanding of how to craft and implement "Business of Humanity" strategies to generate economic value and simultaneously enhance social benefit.

THE BUSINESS OF HUMANITY PROPOSITION

The Business of Humanity (BoH) Proposition is the synthesis of counterintuitive but simple and powerful ideas about how companies can add value in today's globalized and fast-changing world. The fundamental proposition of the BoH is that

> Strategic decision making that employs criteria
> falling under the rubric of "*humanity*,"
> in its two dimensions of "*humaneness*" and "*humankind*,"
> leads to superior economic performance and sustainability.

The BoH Proposition leads to *humane* decisions and actions that will increase profits and, importantly, reduce missteps when entering a new and untested market. The BoH responds to *humankind* in business decisions by recognizing and employing the value-enhancing potential of emerging markets at the bottom of the pyramid.

While it may appear reasonable to presume that our focus on "humanity" has a numinous or moral basis, we must emphasize that the foundations and support for the BoH Proposition are dominantly empirical. The BoH Proposition is based on a combination of inductive logic and empirical evidence. It was formally articulated in 2008 based on theoretical underpinnings and subsequently explored empirically through case studies in four continents. Over the last eight years, we have explored successful, novel, and counterintuitive strategies and innovative business models aligned with the BoH Proposition in companies in Brazil, the Czech Republic, India, Russia, and the United States. We have observed the importance and relevance to strategic decision making of factors that have been traditionally treated as "externalities."

These organizations were selected for study because they were known or perceived to be employing BoH-related business models. Information was gathered from internal documents and published materials and through (mostly videotaped) interviews with two to three senior executives, usually including business unit heads. The organizations that we have studied from a BoH perspective and the locations that were visited by us are listed below.

- Acusis (India and the United States)
- Albert Einstein Hospital (Brazil)
- Alcoa (Russia and the United States)
- Apollo Hospitals (India)
- Arvind (India)
- Bharti Enterprises (India)
- CavinKare (India)
- Coca-Cola (Brazil)
- Dow (Brazil)
- DuPont (Brazil)

- EMCO (Czech Republic)
- Ford (Brazil)
- HCL (India)
- iGate (the United States and India)
- Lilly (Brazil)
- Murugappa Group (India)
- NASSCOM (India)
- RTBI (India)
- Solectron Centrum (India)
- Tenet (India)
- Vodafone (Czech Republic)
- WIPRO (India)

In addition, two organizations that we have not visited, but with which we have become familiar through published sources, have provided us with high confidence that our focus on "humanity" has both social and economic meaning. These two organizations are Aravind Eye Hospital and Unilever. These two organizations powerfully and evocatively demonstrate that "humanity," when effectively integrated into business strategy, is a powerful driver of economic value.

This book is based on these case studies, our consulting experience, our evolving theoretical understanding, and a BoH-based, global strategic initiative that we have undertaken. We offer a commonsense rationale for the power of the BoH Proposition. More importantly, we offer a pragmatic approach—a framework and a process—to enable companies to develop and confidently implement value-adding strategies based on the BoH Proposition.

While the incorporation of BoH into an organization's strategic thinking might seem to be a do-gooder's fantasy, it is not about corporate social responsibility (CSR) for its own sake. It is based on an informed understanding and response to the shortcomings and inadequacies of accounting numbers and, more importantly, the proven synergy of economic and social value-added.

ORIGINS OF THE BoH PROPOSITION

There is no "aha" moment associated with the conceptualization of the BoH. The early stirrings of the idea began in the 1960s. There was a rich body of management thinking that hinted at what we have now formalized—as for example, Professor John Dearden's writings. Professor Dearden (1968, 1969) demonstrated that using return on investment (ROI) as a yardstick for top management to make resource allocation decisions had major limitations and was likely to be seriously dysfunctional. Among ROI's many limitations is its susceptibility to manipulation.

There are many reasons why using projected accounting profits to guide corporate strategic choices is problematic. For starters, accounting profits are not the same as economic profits, as historical costs are taken into account rather than opportunity costs. Moreover, the whole business of full costing itself is open to manipulation owing to the semiarbitrary allocation of costs and the possibility of boosting reported profits by increasing inventory levels when sales are stagnant. Accounting numbers are a shadowy reflection of current economic realities. More fundamental measures of economic health need to be adopted to add substance and meaning to accounting numbers, which, needless to say, are still vitally important from a shareholder's perspective. And, when entering new markets—as we well know but often choose to disregard—projections of profit are often just expressions of hope rather than reality.

When assessing strategies, if you look only at profits, you are relying on a single measure but forgetting all the assumptions and uncertainties that underlie profit projections. You are looking for the comfort of a number built on a foundation of assumptions that are particularly uncertain and questionable in the context of strategic decisions that seek to transform the organization and also change the business context.

In the early 1950s, General Electric's seminal measurements project identified eight key result areas (KRAs) in which every company had to do well in order to be sustainable:

1. Profitability
2. Market Share
3. Productivity
4. Product Quality
5. Management Development
6. Employee Morale
7. Public Responsibility
8. Balance between Short Term and Long Term

RECOGNIZING HUMANE CRITERIA

GE's KRAs provide support for the *"humane"* component of the BoH Proposition. KRAs such as "Public Responsibility," "Employee Morale," "Management Development," and "Balance between Short Term and Long Term" are obviously aligned with the premises underlying the BoH Proposition. GE's measure of "Profitability" also improved on the traditional accounting measure of profit by developing a concept they called "Residual Income." Residual Income moved a step closer to economic profits as it deducted a charge for capital employed from profit after tax. This "income after a capital charge" measure has evolved into the now widely used "economic value added" (EVA) measure.

We have observed that if the detailed design and finer aspects of strategic initiatives are responsive to and aligned with what we term "humane" criteria, EVA can be enhanced. We have found strong evidence supporting the value-adding potential of several "humane" criteria. It is important also to recognize that these criteria address some of the inherent shortcomings of employing accounting profits as the sole criterion for evaluating and shaping transformational strategic initiatives.

The humane criteria we propose are substantially free of many of the shortcomings of accounting profit. For instance, enhancing quality, using the techniques and tools of the Six-Sigma approach, can be counted on to be beneficial to the bottom line. Seeking to enhance environmental sustainability can stimulate innovations in both products and processes that enhance profits much more sustainably than debilitating cost reduction approaches such as downsizing, right sizing, restructuring, or other euphemisms for laying off employees.

Projecting profits in the context of the kinds of uncertainties (Courtney et al. 1997) that businesses face today is a difficult and questionable exercise. The many assumptions that are made, the variety of possible futures that could be encountered, the potential for manipulation and the difficulty of determining cause–effect relationships make projections of profit quite suspect. Employing humane criteria, on the other hand, can result in the development of action plans and programs that are likely to be robust and effective in a variety of future scenarios. We will show you how to develop such robust actions and programs using the simple but powerful possibility-scenario planning process. The intrinsically reinforcing nature of actions that support these humane criteria makes the strategies even more robust.

It is noteworthy that the largest company in the world, the \$480 billion Walmart, has launched a Sustainable Product Index to track the environmental performance of suppliers and has created a President's Global Council of Women Leaders. Meeting the environmental sustainability and gender equality goals that these support affects the way business is conducted by Walmart; they are part of its business model.

These sustainability and equality initiatives are quite different from Walmart's commitment of \$2 billion toward reducing hunger in the United States. Reducing hunger in the United States can be seen as a combination of Walmart's commitment to social responsibility and public relations, demonstrated by allocating a portion of profits to good causes. Reducing hunger is not a strategic goal of Walmart or an intrinsic element of its business model and is not seen as related to its profit goals. In contrast, initiatives such as the Sustainable Product Index and the Global Council of Women Leaders have been described by Walmart's CEO, Mike Duke, as not only the "right thing to do" but, importantly, as also "good for business."* "Good for business" of course means "good for profits."

There is an increasing awareness that paying attention to considerations such as safety, quality, and environmental sustainability engenders dividends for the bottom line. Companies such as Alcoa affirm safety as their unquestioned and primary strategic consideration. These are not just public-relations-oriented statements but accepted and working criteria that affect both strategic and operational decisions and practices. A partial list of such humane criteria includes safety, quality, environmental sustainability, diversity, gender equality, and integrity (Camillus 2009).

HUMANENESS AND ECONOMIC PERFORMANCE

Safety, as Union Carbide discovered subsequent to causing the largest industrial catastrophe in history in Bhopal in 1984, has implications for the survival of the organization. After arrests of executives, appearances before the U.S. Congress, introduction of additional government regulation in the United States and India, and payment of substantial fines, Union Carbide India Limited and the U.S. parent company Union

---

* See: http://www.forbes.com/sites/eco-nomics/2012/04/19/wal-mart-expands-sustainability-initiatives-in-new-report/2/#3e2234e92747

Carbide have been taken over by Eveready Industries in India and by Dow Chemicals in the United States. British Petroleum's multibillion dollar oil spill disaster in 2010 has reaffirmed the importance of giving primacy to safety in managerial decision making. The 2010 Chilean mine disaster, despite the heroic rescue of all 33 trapped miners, has resulted in the closure of the mine and a promise of countrywide government regulations to ensure the safety of workers in industrial, agricultural, and fishing occupations. In the first half of 2014, after political and public outrage at its non-responsiveness to reports of accidents and deaths, GM has recalled more cars than were sold in the United States by all car manufacturers in 2013. The severe economic consequences of not paying attention to safety are, unfortunately, abundantly evident.

Safety has ongoing operational implications for the bottom line in addition to episodic strategic consequences (Oxenburgh et al. 2004). Attention to safety can result in enhanced employee morale and potential efficiency improvements in operating practices, as well as reduced downtime, product liability, fines, jail terms, medical expenses, and regulatory oversight. Disregard of safety is not an option, and attention to safety can help an organization to survive and enhance its profitability. Concern for safety, in the work environment and in product offerings, clearly aligns with organizational values that give importance to employees, customers, and shareholders.

It was logical, therefore, for Paul O'Neill, CEO of Alcoa, to embrace safety as a strategic imperative and the most important criterion for strategic decisions.

Quality has not always received and, even today, does not sometimes receive the unquestioning managerial acceptance that tends to be accorded to safety in public pronouncements. American automobile companies are an interesting example of the evolving approach of businesses to product and service quality. W. Edwards Deming, the now-famed consultant, U.S. government employee, and professor, found little favor in the American industry for his approach to quality enhancement. Even after receiving Japan's Order of the Sacred Treasure in 1960, in recognition of his contributions to reviving Japanese industry and immeasurably improving the Japanese reputation for quality, it took two more decades before the first American automobile company, Ford, thought to employ Deming and implement his approach to quality.

Investments in quality were traditionally seen as negatively affecting the bottom line. It took a wholesale shift of American customers from

American-made automobiles to Japanese-made automobiles for the American automobile industry to rethink its approach to quality. The most recent independent assessments of quality in 2010, by organizations such as Consumers Union and J.D. Power, indicate that American auto manufacturers are now capable of matching and potentially exceeding Japanese quality. Ford has paid more than lip service to its slogan of "Quality is Job 1." In fact, under Ford's management, Jaguar automobiles rose from the bottom of the list in quality and reliability to heading the list in 2009. It is noteworthy that Ford's attention to quality may be one of the factors that enabled it to be the only one of the U.S.' Big Three automakers to avoid declaring bankruptcy or asking for a government bailout.

To attain ISO certification or win Malcolm Baldrige awards, an unwavering focus on quality in processes, services, and products is essential. Meeting the ISO 9001 standard has been shown to result in higher growth rates for sales, employment, payroll, and average annual earnings. ISO 9001 organizations also have a higher likelihood of organizational survival (Levine and Toffel 2010).

In the health arena, with the greater transparency and the growing expectations of clients, an emphasis on quality is crucial. Total Quality Management and Six-Sigma programs are, today, justified on the basis of their ability to enhance revenue and reduce costs, while at the same time enhancing quality. Beyond the fundamental advantage of retaining and attracting customers, it is now common knowledge that well-managed quality programs affect the bottom line immediately, contributing more than they cost (Crosby 1979).

No wonder, Bill Ford, chairman of Ford, has declared quality to be "Job 1."

Environmental sustainability has, for decades, been accepted as a high priority in many European countries. In the last several years, the United States has also come to recognize and increasingly value environmental sustainability. Some of the motivation for this undoubtedly stems from a sense of social responsibility (Marcus and Fremeth 2009). Other reasons might include economic considerations such as the necessity to respond to, for instance, the Restriction of Hazardous Substances and Waste Electrical and Electronic Equipment Directives enforced by the European Union. Without green technology, green processes, and green products, access to the significant European market is limited. Competitors in Asian countries such as China and India may be better situated to meet the regulatory requirements of the European market as their manufacturing plants are newer or being built to green specifications. Manufacturing plants in the United States are more likely to be already in place and be designed and built in an era where environmental considerations were not as urgent and compelling. Nevertheless, though U.S. manufacturing may find it costly to respond to these regulatory directives, the opportunity costs of noncompliance may be too high.

In general, as Krugman (2010) phrased it, "any serious solution must rely mainly on creating a system that gives everyone a self-interested reason...." These self-interested reasons are being fortified by a system of governmental incentives—subsidies and penalties. Krugman claims that economists tend to overestimate the cost of going green and that economic arguments tend, inappropriately, to ignore what they perceive as externalities. There is an emerging argument and realization that environmental social responsibility can enhance profitability. This possibility has emerged because of the system of governmental incentives, better analytical tools, and the more realistic "endogenization" of externalities (Siegel 2009).

Green technology and innovative green products also offer the potential to considerably augment revenues. Green products are expected to provide both growth opportunities and competitive advantage (Unruh and Ettenson 2010). The government subsidies and tax incentives mentioned earlier strengthen the economic motivation to go green. It is not uncommon, as a consequence, to hear the mantra that "green is gold" (Esty and Winston 2006). There is, indeed, growing empirical evidence that strong environmental management is linked to improved financial performance (Klassen and McLaughlin 1996; Russo and Fouts 1997).

It is easy to understand why Pedro Suarez, the highly successful president of Dow Latin America, has affirmed that "sustainability is strategy."[*]

Diversity, in multiple dimensions, beyond just the racial and ethnic backgrounds of employees, is seen as important to economic performance. At the basic level of race and ethnicity, many businesses recognize the importance of a workforce that can identify with and communicate with potential, targeted customers. A Hispanic radio station without Latino employees would find it hard to function effectively. An American multinational without Asians in its management ranks, with global headquarters situated in a small Midwestern town,

---

[*] See: https://www.youtube.com/watch?v=hvuJp1hyD3Y&feature=youtu.be

might find itself disadvantaged when entering a Pacific Rim nation. So, it is evident that a lack of diversity clearly has bottom-line implications in certain contexts. Companies as varied as Timken and Google value diverse workforces as a source of innovation and long-term economic sustainability.

In other contexts, the argument is not necessarily as clear-cut. The argument is sometimes made that diversity can negatively affect the bottom line because of more conflicts and less cohesion in the workforce (Herring 2009). However, Herring's empirical study supports the contention that diversity benefits the bottom line, across the board. It has been shown that racial diversity is associated with greater profits, greater revenues, more customers, and greater market share.

While the reasons for and directionality of the connection between profits and racial diversity can be debated, situations where the connections between diversity and profits and the directionality from diversity to profits are clear are not uncommon. There are also penalties and civil liabilities for engaging in racial and other forms of discrimination that might motivate organizations in their own best economic interests to support diversity.

Gender equality in wages and job and career opportunities is a long-standing women's issue that is perceived to be complex and contradictory in terms of economic impact on businesses and nations (Richard et al. 2004; Sequino 2000). In this somewhat murky vein, Blau et al. (2006) provide a data-driven set of somewhat conflicting perspectives on this issue.

Gender inequality, however, is widely viewed as unacceptable. Nations that constrain women's education, opportunities, and rights are seen as losing entirely or reducing considerably the potential economic and cultural contributions of half their populations. Gender inequality has declined and is expected to continue to decline because of women having better access to education, the changing nature of work, the shift from blue-collar to white-collar work, more equitable sharing of household responsibilities between genders, better child-care options, enhanced maternity benefits and parental leave, and more enlightened societal mores and legislation. Walmart, for instance, faced a class action discrimination lawsuit, originally brought by six women employees, that could have cost the company billions of dollars (O'Keefe 2010). The U.S. Supreme Court, in a divided decision, acquitted Walmart, but the issue and potential liability are not likely to go away.

While there are continuing debates about the economic consequences of the gender gap in pay, a negative effect on women's morale and identification with the organization seems inevitable (Gonzalez and DeNisi 2009; Leete 2000). On the positive side, Herring's (2009) study found that gender diversity is associated with greater revenues, more customers, and higher profits. Campbell and Minguez-Vera (2008) found less comprehensive but similar results regarding the financial impact of gender diversity. If gender diversity is seen to have a positive impact on the bottom line, it can reasonably be argued that gender equality has positive profit impacts on businesses (Dickens 2006).

Integrity is similar to safety in that its importance for organizational survival is self-evident. The concept of integrity is intrinsically humane in that it defines the desirable character of the interaction between individuals. Integrity leads to interpersonal respect and enhances the quality of life. Becker (2009) also argues that integrity is a prized company asset as it enhances reputational capital and the ability to differentiate.

Michael Jensen, the doyen of finance professors in the 1990s, goes further and posits a link between integrity and financial performance. According to him, "integrity is a necessary condition for maximum performance" (Christensen 2009). Erhard et al. (2008) demonstrate that integrity, honoring one's word as an individual or a group, has a causal link to increased performance. They further demonstrate that integrity is humane in that it also leads to improvement in the quality of life.

HUMANKIND AND ECONOMIC PERFORMANCE

In today's global economy, recognizing the universe of market possibilities, it makes sense to consider "*humankind.*" Market growth, through both expansion and new markets, takes place today in emerging economies such as Brazil, Russia, India, and China—the BRIC countries. Also, as the "fortune at the bottom of the pyramid" (Prahalad 2006) suggests, there is a strong economic incentive to address the needs of low-income segments because of their incredibly large, $13 trillion of purchasing power.

The rapidly globalizing business environment has seen corporations from the developed world invest around $3 trillion in emerging economies during the last two decades—a corporate-investment cycle that is one of the biggest since the railway boom of the nineteenth century, according to the *Economist* (2014) magazine. This boom dates to a period when India deregulated, decontrolled, and delicensed its *dirigiste* economy in the 1990s and China's liberalized economic policies ignited economic growth. By the mid-2000s, Brazil, Russia, India, and China* caught hold of the imagination as future engines of growth as corporations piled into these economies.

This massive corporate-investment wave is now fast breaking. To be sure, there are winners and losers in this process. The companies that have done well have been cognizant that there are billions of people at or close to the bottom of the pyramid in the emerging and BRIC economies who constitute a market no less important than the affluent middle class. The demographics of these economies are also different with a growing population of youth, which will result in the dividend of a significant workforce. These ranks will soon acquire greater purchasing power and market influence, breaking into the lower rungs of the middle class and motivating disruptive and frugal innovations in business models.

For instance, when a company tries to meet the needs of impoverished people in large numbers, it forces a rethink of products and services, technology, and marketing—in other words, the development of innovative business models. How would the company design, make, sell, and deliver the products? The technologies needed and the processes used to answer such questions are enormously powerful motivators for innovation (Hart and Christensen 2002). The disruptive innovations stimulated by a globalized world heighten the challenges and importance of business's role in society. The backlash against globalization, which is often based on political motivations rather than on macroeconomic understanding, makes the role of corporations even more challenging.

To cope with pressure from anti-globalization movements, more and more transnational corporations (TNCs) are forced to adopt strategies that go beyond short-term profit maximization. One such strategy is to embrace social business enterprise—a fuzzy idea that doesn't admit of a precise definition but which broadly conflates the interests of TNCs with socioeconomic development. TNCs are taking this approach in order to cope with the growing discontent against globalization, which is perceived

---

* The BRIC acronym came into fashion 17 years ago when a Goldman Sachs report held out India's role as a major player along with Brazil, Russia, and China.

to benefit only the affluent segments of the developed world rather than those at the bottom of the pyramid.

But a fuzzy idea of "doing well by doing good" often needs a prophet more than erudite management gurus. Groupe Danone found that in Nobel laureate Muhammad Yunus, the father of microcredit. Yunus met with Franck Riboud, Group CEO of Groupe Danone in November 2005 and persuaded the French giant to set up a yogurt factory in Bangladesh as a joint venture with Grameen Bank, Grameen Danone Foods Ltd. This yogurt is fortified with vitamin A, zinc, and iodine to help combat malnutrition and sold at an affordable rate of six euro cents a cup. The factory relies on Grameen microborrowers buying cows to sell it milk and microvendors selling the curds from door to door.

What is in it for Danone? Certainly not just the satisfaction of providing charity. This venture actually makes strong business sense as it creates economic and social dividends for its shareholders. Imagine the company reporting in its annual statement the number of children that it saved from malnutrition, improvements in public health, employment creation, poverty reduction and environment protection! Imagine a world in which more and more transnationals like Danone raised such "social capital" and invested it in sustainable businesses in poorer developing economies like Bangladesh and sub-Saharan Africa?

It must be noted that Danone, of course, was predisposed to the call of prophets of social responsibility. For instance, Danone has also been providing safe drinking water in India since 2010 through the Naandi Foundation. The Naandi Community Water Services is present in 400 villages, providing safe drinking water to 2.4 million people. Here, again, the effort is to provide it at an affordable rate of 0.3 U.S. cents per liter. The program runs on a hybrid model, with governments and investors contributing 70% of its initial capital. Now, it is managed as a business that incorporates social benefit in its business model to expand the number of treatment units from 500 to 2000 sites. Danone assists with issues related to quality, marketing, human resources, the program's supply chain, and IT.

It is our intention to demonstrate to companies and executives, who are not predisposed the way Danone was, that the BoH Proposition is sound and should be embraced as conventional wisdom. In fact, it is our thesis that utilizing BoH principles will significantly increase the probability of success in new and emerging markets. We will illustrate this by offering both logic and compelling examples throughout the book.

INTEGRATING HUMANENESS AND HUMANKIND

The BoH Proposition views "Humanity" as an integrating construct that incorporates "Humankind" and "Humaneness." Humankind serves to stimulate the identification of innovative and transformational strategic alternatives. Humaneness supplies the criteria for evaluating these strategic alternatives. Humanity as the construct that integrates humankind and humaneness, with a supporting set of organizations that illustrate the importance of the elements, is diagrammed in Figure 1.1. The companies, which are illustrative of humane values or of responsiveness to humankind, that are listed below each set of values or responses have been described earlier in this chapter or are discussed following Figure 1.1.

LESSONS FROM THE REAL WORLD

Examples of the power of the components of "humanity"—humaneness and humankind—are numerous. The power and potential of emerging markets to support disruptive innovation and grow profits is strikingly

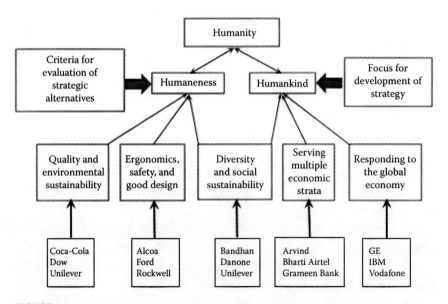

FIGURE 1.1
Humanity as an integrating construct.

evidenced by *GE*. Jeffrey Immelt, CEO of GE, shifted the development of medical imaging technologies and devices—low-cost electrocardiogram and ultrasound machines—to China and India, with innovation efforts directed toward **meeting the needs of their vast low-income markets**. The hoped-for "reverse innovation" has worked brilliantly. GE's innovation initiatives have delivered hi-tech solutions at ultralow cost. GE's **frugal engineering** capability enables it to achieve a position of cost leadership. The ultralow-cost position empowers GE to access the base-of-the-pyramid markets. It also enables GE to "preempt" local companies in these emerging countries from creating similar products to disrupt GE in rich countries (Immelt et al. 2009). GE's strong and growing emphasis on growth through focusing on water and infrastructure is a further illustration of the relevance of the BoH approach in strategy—**meeting fundamental human needs**, with emphasis on the base of the pyramid.

Walmart's positive and negative experiences over the last decade are strong evidence of the validity of the BoH Proposition. Its efforts to **address environmental concerns and sustainability** in the area of packaging led to savings of billions of dollars, which motivated a broader approach including **reducing the carbon footprint** of its suppliers.

To work within the regulatory framework of the Indian government, Walmart had to create new business models using its existing competencies. Because of the Indian government's ban on multibrand retailers entering its retail sector and displacing millions of small shopkeepers, the world's largest retailer perforce has had to **gain social sustainability** by operating as a wholesaler, serving local retailers with cash and carry stores.

On the negative side, Walmart has given up on or indefinitely postponed its entry into the retail sector in India by deciding not to undertake the admittedly rigorous requirements placed on foreign multibrand retailers by the government. In Walmart's estimation, these are onerous requirements that would cripple its profit-making capacity. Local sourcing requirements of 30% (Walmart was only willing to commit to 20%) may have seemed onerous, but our observations, bolstered by Ikea's experience in India, has suggested otherwise. Local sourcing could have been readily extended beyond the Indian market to serve Walmart's global market.

The governmental requirements could actually have proved to be a source of opportunity, of more directly and comprehensively addressing the needs of India's hundreds of millions of poor people. Walmart has missed a game-changing opportunity in this regard. Investing in the back-end infrastructure need for setting up a cold chain could have ushered in a farm-to-fork revolution, and Walmart preemptively, in combination with its wholesale operations, could have captured an unassailable competitive position and unparalleled social sustainability. Millions of farmers would have benefited in the process. Walmart's entry into India arguably might have had a different and more profitable outcome had it wholeheartedly **embraced the BoH proposition**.

Eleven years ago, *IBM* CEO Sam Palmisano called on traditional multinational corporations (MNCs) to abandon their "colonial approach" to operations outside their home country to avoid an anti-globalization backlash. What he had in mind were the Big Three car giants—GM, Ford, and Chrysler—that built factories in Europe and Asia but kept their R&D back home. And then there were glocalization strategies of companies like GE, which used to develop high-end products in developed markets and marginally adapt them for emerging markets around the world. These MNCs needed to **move toward full global integration of their operations** to be closer to their customers.

Bharti Enterprises, which is the conglomerate that owns Bharti Airtel, India's largest private cell operator, is very much a globally integrated enterprise as defined by Palmisano. There are reports of it bringing more services back in-house as a result of customer complaints of poor service. According to an article ("Business is going native again") by Andrew Hill (*Financial Times* 2014), the challenge is to **maintain trust**, both internally and externally. As such issues have come to the fore, the biggest beneficiary has been the customer, with perhaps the lowest mobile phone tariffs in the world (less than one U.S. cent a minute) and a choice of technology—GSM or CDMA. India now has more than a billion phones, 98% of which are mobiles.

For a 1.2 billion-person nation, such penetration rates make for a fascinating case study for the BoH. There are currently 46 mobiles for every landline phone, as the revolution in mobile telephony has gone beyond the big metropolises, smaller cities, and towns to **penetrate the impoverished nooks and crannies of rural India**—enabling farmers, for instance, to know prices in neighboring markets before they off-load their produce. Similarly, fishermen in the State of Kerala use mobiles to check fish prices in order to land their catch ashore at the most favorable harbor. The ongoing telecom boom is **socially inclusive** in nature as it helps bridge the divide between elite and rural India.

A striking example is Ahmedabad-based *Arvind Limited*, the flagship of the Lalbhai Group in India. Arvind's revenues were INR 55 billion (rupees) or US$843 million in 2015–2016. The company, which is one of the world's leading producers of denim, is currently run by the fourth and fifth generation of its founders, Sanjay Lalbhai and his two sons, Punit and Kulin.

Arvind has long dedicated part of its profits to supporting social causes. Its exercises in social responsibility include entrepreneurial training for widows, improving local schools, providing vocational training for the rural poor, and providing clean drinking and better sanitation in the slums of Ahmedabad. These commendable activities fall into the realm of traditional CSR practices.

In recent years, Arvind has gone beyond traditional, post–profit-making CSR. Arvind's recent **CSR efforts are incorporated into business models** that are innovatively crafted to simultaneously address social issues and enhance profits. A brilliant example is its effort to address an epidemic of suicides by cotton farmers in the nearby Vidarbha region. It developed a business model that redressed the destitute condition of cotton farmers working miniscule patches of nonirrigated land and simultaneously grew the company's competitive advantage in organic cotton denim.

Arvind purchases cotton from farmers for its textile operations. Rain-fed agriculture has its own risks for farmers who typically borrow money at usurious rates of interest for seeds, fertilizers, and pesticides to cultivate cotton. If the rains fail, the crops wither and farmers face the loss of their lands to the moneylenders. Knowing nothing else but farming, landless farmers face starvation. In large and increasing numbers, they choose suicide.

To redress this narrative of farmer distress, Arvind hired the services of agronomists to teach farmers about organic farming or growing cotton naturally, which obviates the need for fertilizers and pesticides. Organic farming is also best suited for marginal farms with poor soil quality as is prevalent in Vidarbha. What started out as a small project now encompasses 41,471 acres of farm land and 5482 farmers. All the organic cotton produced on these farms is certified by Control Union Certification, Netherlands. Arvind has eliminated middlemen and directly picks up the farmers' production. The company has negotiated commitments from global companies to purchase the organic cotton at a premium price. Previously **indigent farmers now make a good living, suicides in Vidarbha have dropped, the soil is enriched, the environment is protected**, and Arvind has become the largest and most profitable manufacturer of organic denim.

A video of an interview with Sanjay Lalbhai, chairman and CEO of Arvind, is available.* It adds detail and authenticity to our discussion.

Ford Motor Company, like other global corporations who were attracted by the opening up of India's vast market since the early 1990s, initially introduced products that were targeted for international acceptance. Ford, thus, kicked off its presence in India with its Escort model that was a winner in Europe. However, like others, it soon realized that the market at the upper end was narrow; that it had to develop **products targeted to low-income consumers** to stay in business. The low-cost Ikon was developed and the rest, as they say, is history. Ford has major expansion plans in India.

Ford is of particular interest to the BoH project as it is the only one among its U.S. rivals to have survived the global recession of 2008–2009 intact, without filing for bankruptcy or relying on government intervention. As mentioned previously, its cars also match the reliability of Japanese cars and its hybrids are among the best in the world. The humane ideals the company has embraced can be seen to have had a role in creating uniquely successful performance. Witness the much advertised jingle that **quality is "Job 1"** and Ford's assigning **mission-critical importance to sustainability**. Furthermore, Ford's emphasis on **gender equality** has resulted in the following:

- Product designs that are responsive to women's physical characteristics. One example of such thoughtful and responsive design is the introduction of moveable pedals to reduce injuries to women from the deployment of airbags, caused by having to sit too close to the steering wheel.
- Its Camaçari facility located in a backward region of Brazil, with the highest percentage of women on the assembly line, became one of the world's most advanced, environmentally friendly, flexible, and productive car manufacturing operations.

Ford's Camaçari plant with its **emphasis on the environment**—being, for example, the first major plant in Brazil to use wetlands instead of chemicals for treatment of sanitary waste—has found that its profits have been enhanced as a consequence.

But if there is one poster child for the BoH Proposition—especially with respect to the potential of markets at the bottom of the pyramid—it surely is microfinance. Extending small sums of credit is an important

* See: https://www.youtube.com/watch?v=VPC8Kn_XaSA

way to reach out to the unbanked population in rural India. Providing microcredit particularly to women, to purchase, say, a sewing machine or a vehicle to transport goods to the urban markets, is a proven development strategy—demonstrated by the spectacular success of the **Grameen Bank**, established 34 years ago in Bangladesh. The inspiring saga of Grameen Bank triggered a global microfinancing movement, **enabling the very poor to become entrepreneurs**.

Naturally, the good news spread across to India, thanks to the influence of multilateral agencies and nongovernmental organizations (NGOs). Provision of modest loans to rural women has almost always resulted in 100% repayment rates. The collateral often is only the commitment of the community or self-help groups (SHGs). Once they are started, they rapidly become self-sustaining in nature, rarely requiring periodic grants and similar forms of assistance. SHGs now constitute 19 million or 6% of working-age Indians. The growth and success of the microfinance industry have also attracted the attention of private equity and hedge funds.

The contrasting fortunes of SKS Microfinance Private Ltd and Bandhan Financial Services Private Ltd, however, highlight what can happen to a good idea if firms lose sight of larger societal imperatives. The saga of **SKS** is a sorry tale of a company that lost its way and reverted to traditional strategies in an industry crying for the application of the BoH approach. Not so long ago, Vikram Akula, founder and CEO of SKS, was feted as a leading social entrepreneur. Based in Hyderabad in Andhra Pradesh, SKS applied global business practices to the field of microfinance. The company was hailed as the "Starbucks of Microfinance" for standardizing microfinance processes and using technology to accelerate growth. It had a massive loan book of $1.2 billion, one-third of which was in Andhra Pradesh, when it went public in July 2010.

The successful IPO listing signaled the coming of age of microfinance in the India scene. But shortly thereafter, it started to unravel for SKS and the industry in general. What originally started as an NGO movement to meet the credit needs of poor rural women lost sight of this BoH-related objective in the go-go years; SKS started forcing multiple loans on borrowers at usurious rates of 30% or more. Not surprisingly, the government of Andhra State cracked down on the industry for its strong-arm methods in enforcing collections that drove farmers to suicide. Almost all the loans were written off. SKS bore the full brunt of the government's ire. Its loan book shriveled to $325 million.

Contrast SKS Microfinance's experience with that of Bandhan Financial Services Private Ltd. **Bandhan** started in 2001 under the leadership of Chandra Shekhar Ghosh to work for the **uplift of socially disadvantaged and economically exploited women** in East India. In 12 years, this institution has extended its sphere of operations to 22 states, with a loan portfolio of $1.5 billion, offering reasonable interest rates to 6.5 million borrowers. Bandhan's socially responsible business model has even led to India's central bank, Reserve Bank of India, granting it a rare and prized banking license—the first to a microfinance institution.

The award of a banking license to Bandhan clearly is another high for the microfinance industry. It will now be in a position to tap rural savings and bolster lending to the unbanked segments of society. Just 53% of the country's adult population has bank accounts. With access to a larger pool of capital, Bandhan will be in a better position to lower its borrowing costs and charge even more reasonable rates of interest. Such initiatives clearly make for a more **inclusive** form of providing financial services. Banking on the unbanked is another big idea whose time has come, if microfinance fulfills its promise of meeting the needs at the bottom of the pyramid.

This tale of two companies is compelling. Both operated in the same well-defined industry, in the same country (India), and both started with a commitment to a strategy aligned with the BoH Proposition. SKS Microfinance, in preparation for an IPO, abandoned humane practices in the pursuit of profits. Bandhan, in contrast, remained true to its original values and BoH strategy. SKS Microfinance stumbled badly, but Bandhan was rewarded for its adherence to its ideals with the banking license from the Indian government, which is analogous to finding the Golden Fleece. The BoH strategy triumphed.

The instances of BoH principles being integrated into business models are seemingly endless. An example of MNCs following such strategies is *Coca-Cola*, which has set for itself an aspirational goal of **becoming water-neutral in its operations on a global basis**. This means returning water used in the manufacturing process back to the environment in a form that sustains aquatic life and helps protect watersheds where the company operates. According to *Rockwell*'s famed CEO of the last decades of the twentieth century, Donald Beall, the evolution and transformation of Rockwell and the superior returns that it continues to generate stem in large part from the **culture and character of the company that are derived from the humane, core beliefs of its people.** *Alcoa*, which under CEO Paul O'Neill made **safety its highest strategic priority**, turned around its acquisitions in Russia by employing safety as the focus of its strategic decision making.

Vodafone's unit in the Czech Republic is highly profitable and one of this global company's crown jewels in terms of culture and employee loyalty. Interestingly, **in its strategic plan, profits are mentioned only at the fourth level in the hierarchy of enablers for success.** At the highest level, the focus is on "**breaking all the rules for the customer.**" Customer service representatives have the ability to rewrite company policy on the spot to solve problems for the customers with whom they are speaking. As a result, 2000 revisions to company policy were made in three years! And, counterintuitively, profits are high.

Unilever's chairman, Paul Polman, in the May 2014 issue of *McKinsey Quarterly* says, "For instance, one of our targets is **creating new jobs for 500,000 additional small farmers. ...** The same is true for moving to **sustainable sourcing** or **reaching millions with our efforts to improve their health and well-being.** All of this is hardwired to our brands and all our growth drivers." In line with the BoH Proposition, he adds, "So we created the Unilever Sustainable Living Plan, which basically says that we will **double our turnover, reduce our absolute environmental impact, and increase our positive social impact. ...** That's very motivational for our employees. ... makes the difference between a good company and a great one."

Let us briefly highlight the lessons drawn from these many examples.

Alcoa:
- Make safety the highest strategic priority.

Arvind:
- Recognize the difference from traditional CSR.
- Incorporate social causes in the business model.
- Help the indigent make a good living.
- Protect the environment.

Bandhan Financial Services:
- Uplift socially disadvantaged and economically exploited women.

Bharti Enterprises:
- Maintain trust.
- Penetrate the impoverished nooks and crannies in the populace.

Danone:
- Embrace social responsibility to do well by doing good.

Coca-Cola:
- Become globally water-neutral.

Ford:
- Make quality "Job 1."
- Elevate sustainability to "mission-critical" status.
- Emphasize gender equality.

GE:
- Meet the needs of their vast low-income markets.
- Meet fundamental human needs.

Grameen Bank:
- Enable the very poor to become entrepreneurs.

IBM:
- Move toward full global integration of operations.

SKS Microfinance:
- Do not abandon BoH once initiated.

Rockwell:
- Derive the culture and character of the company from the humane, core beliefs of its people.

Unilever:
- Commit to sustainable sourcing.
- Improve health and well-being.
- Reduce environmental impact.
- Increase positive social impact.

Vodafone:
- Give strategic priority to humaneness.
- Break all the rules for the customer.

Walmart:
- Address environmental concerns and sustainability.
- Reduce the carbon footprint.

- Gain social sustainability.
- Embrace the BoH Proposition.

And the beat goes on among successful companies worldwide: In Japan, **Hitachi**, the global giant, proclaims social innovation to be the basis for building its future; in the Czech Republic, **EMCO**, the small but brilliantly successful Czech company, has a laser focus on the health and nutrition needs of its customer; in Latin America, **DOW** proclaims that sustainability is strategy; and in the United States, **Starbucks** espouses inspiring and nurturing the human spirit as its vision, complementing **Whole Foods**'s statement of higher purpose, "With great courage, integrity and love—we embrace our responsibility to co-create a world where each of us, our communities and our planet can flourish."

These are not idle pronouncements intended merely to impress the naïve and the gullible. These are words and principles that these remarkable companies live by.

The ubiquity of examples of the acceptance, the importance, and indeed the power of elements of the BoH approach warrants and motivates the search for an integrating management framework and management processes, which pull the various strands and elements together, aggregating and enhancing their potential and impact. The management framework should inspire, guide, and create a symbiotic energy, and the management processes need to be laser-focused on building BoH Strategies. Constructing the BoH management framework and designing the supporting management processes are what we endeavor to accomplish in this book.

THE BoH AS THE DOMINANT STRATEGIC MINDSET

Despite the host of examples and the appeal of the underlying logic, spreading the message of the BoH may indeed be a major challenge because strategies driven by the BoH Proposition are very different from what may be suggested by today's dominant strategic mindset. As we will elaborate in the following chapters, the BoH Proposition and construct demand a new way of thinking, a mindset where values infuse and inspire the organization with purpose. As Donald Beall (2008), Chairman Emeritus of Rockwell, argued: people are the organization; people's values determine how people relate to one another and influence decisions in organizations. It is natural that social considerations should be as much a driver of strategic decision making as economic value added.

This is not just about applying greater attention to social responsibility or donating to good causes when distributions of profit are made. **The BoH Proposition mandates that social benefit be incorporated into the corporation's *raison d'être*.**

Of course, the saving grace, the counterintuitive possibility that this book champions is that BoH Strategies—though requiring more insight, commitment, and perseverance than classic strategies solely focused on enhancing shareholder value—do offer the promise of enhancing EVA and sustainability.

The contrast between traditional strategic thinking and the BoH approach is illustrated by the two strategic logics presented in Figure 1.2. Traditional or classic strategic thinking would start with glocalization—minimally modifying products that were developed in advanced economies for customers with substantial disposable income—to respond to the characteristics of upscale markets in the less-developed economies. This would offer the potential of adding to the firm's profits but the additional profits would be limited by the small numbers of customers who have the capacity to buy the products—consider Bentleys and Jimmy Choo—originally developed for and offered to well-to-do customers in the richer countries.

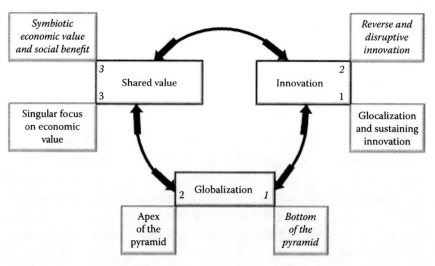

FIGURE 1.2
Classic and BoH Strategies.

In contrast, typical BoH thinking would begin with the goal of identifying community and human needs, recognizing the opportunity presented by the needs of the huge poor and emerging middle-class populations in growing but less-developed economies. This would require disruptive innovation and frugal engineering to meet the dramatically lower, viable price points in these markets. The four billion relatively poor people, with a combined purchasing power that has been estimated to lie between US$5 trillion and $13 trillion, offer the prospect of substantial profits if their needs are creatively met, despite the limitations of their per-capita disposable income.

Moreover, the profit potential that can be tapped goes beyond the emerging economies. Companies that we have studied have shown that the disruptive innovation and frugal engineering underlying BoH Strategies combine to generate what has been described as "reverse innovation" (Govindarajan and Trimble 2012), which offers the attractive possibility of enhanced competitive advantage and greater profits in developed markets.

BoH Strategies—which this book endeavors to show you how to develop— offer the promise of enhanced and synergistic economic and social benefit.

In the following chapters, we will explore the logic underlying the BoH Proposition and offer a framework and processes for confidently developing strategies aligned with the BoH.

REFERENCES

Beall, D.R. (2008). *Formation, Evolution and Transformation of Rockwell*. Newport Beach, CA: Dartbrook Partners.

Becker, G.K. (2009). Integrity as moral ideal and business benchmark. *Journal of International Business Ethics*, 2(2): 70–84.

Blau, F.D., Brinton, M.C., and Grusky, D.B. (Eds.). (2006). *The Declining Significance of Gender?* New York: Russell Sage Foundation.

Camillus, J.C. (2009). Good business in bad times: The strategic advantages of humanity in business decisions, *Effective Executive*, August, 29–34.

Campbell, K., and Minguez-Vera, A. (2008). Gender diversity in the boardroom and firm financial performance. *Journal of Business Ethics*, 83: 435–451.

Christensen, K. (2009). Integrity: Without it nothing works. *Rotman Magazine*, Fall: 16–20, 18.

Courtney, H., Kirkland, J., and Viguerie, P. (1997). Strategy under uncertainty. *Harvard Business Review*, 75(6): 67–79.

Crosby, R.M. (1979). *Quality Is Free*. New York: McGraw-Hill.

Dearden, J. (1968). Appraising profit center managers. *Harvard Business Review*, 46(3): 80–87.

Dearden, J. (1969). The case against ROI control. *Harvard Business Review*, 47(3): 124–135.

Dickens, L. (2006). Re-regulation for gender equality: From "either/or" to "both." *Industrial Relations Journal*, 37: 299–309.

Economist, "Business in Emerging Markets: Emerge, Splurge, Purge," March 8, 2014.

Erhard, W., Jensen, M.C., and Zaffron, S. (2008). Integrity: A positive model that incorporates the normative phenomena of morality, ethics and legality. *Working Paper No. 06-11*. Boston: Harvard Business School Negotiations, Organizations and Markets Unit.

Esty, D.C., and Winston, A.S. (2006). *Green to Gold: How Smart Companies Use Environmental Strategy to Innovate, Create Value, and Build Competitive Advantage*. New Haven, CT: Yale University Press.

Financial Times, April 9, 2014.

Gonzalez, J.A., and DeNisi, A.S. (2009). Cross-level effects of demography and diversity climate on organizational attachment and firm effectiveness. *Journal of Organizational Behavior*, 30(1): 21–40.

Govindarajan, V., and Trimble, C. (2012). *Reverse Innovation: Create Far from Home, Win Everywhere*. Boston, MA: Harvard Business Review Press.

Hart, S.L., and Christensen, C.M. (2002). The great leap: Driving innovation from the base of the pyramid. *Sloan Management Review*, 44(1): 51–56.

Herring, C. (2009). Does diversity pay?: Race, gender, and the business case for diversity. *American Sociological Review*, 74(2): 208–224.

Immelt, J., Govindarajan, V., and Trimble, C. (2009). How GE is disrupting itself. *Harvard Business Review*, 87(10): 56–65.

Klassen, R.D., and McLaughlin, C.P. (1996). The impact of environmental management on firm performance. *Management Science*, 42(8): 1199–1214.

Krugman, P. (2010). Green economics. *The New York Times Magazine*. April 11: p. 39.

Leete, L. (2000). Wage equity and employee motivation in nonprofit and for-profit organizations. *Journal of Economic Behavior & Organization*, 43: 423–446.

Levine, D.I., and Toffel, M.W. (2010). Quality management and job quality: How the ISO 9001 standard for quality management systems affects employees and employers. *Management Science*, 56: 978–996.

Marcus, A.A., and Fremeth, R.A. (2009). Green management matters regardless. *Academy of Management Perspectives*, 23(3): 17–26.

O'Keefe, B. (2010). Meet the CEO of the biggest company on earth. *Fortune*, 162(5): 80–94.

Oxenburgh, M., Marlow, P., and Oxenburgh, A. (2004). *Increasing Productivity and Profit through Health and Safety: The Financial Returns from a Safe Working Environment*. Boca Raton, FL: CRC Press.

Prahalad, C.K. (2006). *The Fortune at the Bottom of the Pyramid: Eradicating Poverty through Profits*. Upper Saddle River, NJ: Wharton School Publishing.

Richard, O.C., Barnett, T., Dwyer, S., and Chadwick, K. (2004). Cultural diversity in management, firm performance, and the moderating role of entrepreneurial orientation dimensions. *Academy of Management Journal*, 47(2): 255–266.

Russo, M.V., and Fouts, P.A. (1997). A resource-based perspective on corporate environmental performance and profitability. *Academy of Management Journal*, 40(3): 534–559.

Sequino, S. (2000). Gender inequality and economic growth: A cross-country analysis. *World Development*, 28: 1211–1230.

Siegel, D.S. (2009). Green management matters only if it yields more green: An economic/strategic perspective. *Academy of Management Perspectives*, 23(3): 5–16. Also: Martin, R.L. (2002). The virtue matrix: Calculating the return on corporate responsibility. *Harvard Business Review*, 80(3): 69–75.

Unruh, G., and Ettenson, R. (2010). Growing green: Three smart paths to developing sustainable products. *Harvard Business Review*, 88(6): 94–100.

2

The Strategic Challenge

There are growing challenges in the business environment that make the Business of Humanity (BoH) approach a logical and indeed necessary response. To demonstrate this, let us take a step back and examine the task facing managers.

The fundamental challenge facing managers is the uncertainty and complexity that characterize business environments. If uncertainty and complexity were not present, there would be no need for managers. In today's business environment, there are powerful forces in play that exacerbate complexity and accentuate uncertainty. These include the following:

- Arbitrage possibilities (especially across national boundaries)
- Technological disruptions
- Regulatory changes
- Political transformations
- Demographic shifts
- Increasingly active stakeholders and nongovernmental organizations
- Changing models of value creation and appropriation

While there are many such factors, in our experience, there are three that are ubiquitous and that, individually and in combination, create extremes of complexity and uncertainty that managers have to face. In learning to respond to these three forces, businesses will find themselves prepared and equipped to handle the other environmental forces. These three forces, which we have found to be dominant aspects and drivers of the business environment, deserve to be labeled "mega-forces." They are as follows:

1. The inevitability and growing importance of *globalization*
2. The necessity and disruptive nature of *innovation*
3. The societal pressure and increasing expectation for *shared value*

As we mentioned, each of these mega-forces* is individually important, but it is necessary to recognize that their interactions also contribute to greater complexity and uncertainty in the environment. We will examine globalization, innovation, and shared value individually before exploring the consequences of their interactions. We will then proceed to explore and explain why and how BoH Strategies respond to the challenges that are born of the interactions between these mega-forces.

GLOBALIZATION

Globalization is an *inevitable*, ubiquitous, and accelerating trend. Even organizations that make a deliberate choice to limit the geographic scope of their target markets will inevitably encounter issues of globalization because of the ever-increasing importance of global sourcing, global competition, global standards, global quality expectations, global partnerships, and global financing. The global connections between firms, in material, human, and competitive terms, are growing in both diversity and intensity.

This growing interconnectedness between firms is reflected in the relationships between national economies. The attractions of both cost arbitrage and vast, growing markets in emerging economies are accelerating the trend toward increasing global connections. The global electronics contract manufacturing industry, for instance, which generated

---

* These mega-forces are discussed in depth in *Wicked Strategies: How Companies Conquer Complexity and Confound Competitors* (John C. Camillus, University of Toronto Press, 2016). "Wicked Strategies" are analogous to but different from the BoH strategies that are offered here. Wicked Strategies are designed to deal with wicked problems resulting from the extreme complexity and uncertainty, the chaotic ambiguity created by these mega-forces. BoH strategies are also designed to address complexity and uncertainty, but they go beyond the generalized approach of Wicked Strategies. BoH strategies are uniquely and distinctively crafted to create and exploit the synergy between economic value and social benefit. They are consciously and carefully aligned, as we will demonstrate, with the trajectory of evolving value systems in most cultures across the world. The BoH Management Framework and the supporting processes are therefore substantially different from the framework and processes that support Wicked Strategies, though the environmental forces that create the need for both Wicked Strategies and BoH Strategies are the same.

an estimated revenue of $360 billion in 2011, was projected to grow to $426 billion by 2015.* While the outsourcing phenomenon was initially motivated by lower manufacturing costs, the reasons supporting outsourcing extend far beyond just lower costs. Today, manufacturing capabilities and skills available in some of the emerging countries exceed those of the developed countries. For advanced digital-timing chips, Solectron Centum in Bangalore, India, is one of the best sources. For automotive forgings, Bharat Forge, a large Indian company, offers both the most advanced capabilities and the highest quality.

In 2008, M.M. Murugappan,† one of the most insightful and successful CEOs we have met, said something which is important in understanding why offshoring is going to accelerate. Speaking about his operations in other countries, he said that he would never consider offshoring just to reduce costs. He said the difficulties and downsides of seeking cost reduction through outsourcing are substantial and the benefits are ephemeral. He emphatically asserted that the only reason he would offshore operations was to access knowledge and skills or markets. This leads us to perhaps the most powerful force, making increasing globalization inevitable—the pull of large and growing markets outside the developed economies.

From their inception, going global was the only option for companies such as Philips and Nokia. Philips, originating in the tiny town of Eindhoven in the Netherlands, would have collapsed quickly if it had limited itself to the Dutch boundaries. The Dutch heritage of exploration and centuries of colonial experience no doubt contributed to its becoming global. For four decades until the 1980s, Philips was the largest and most successful consumer electronics company in the world.

Nokia applied a similar mindset to expand beyond the limitations of the boundaries of Finland and become the dominant cell phone producer in the world. The name Nokia became synonymous with cell phones in the 1990s and continued as such until Apple stunned the world with its iPhone in 2007.

Globalization was a key to success in the case of both Philips and Nokia. But the challenges of globalization also brought about their falls from grace.

---

* http://www.americanprogress.org/issues/labor/news/2012/07/09/11898/5-facts-about-overseas-outsourcing/
† M.M. Murugappan is the vice chairman of the Murugappa Corporate Board. He is the chairman of Tube Investments of India Ltd, Carborundum Universal Ltd, Wendt India Ltd, and Murugappa Morgan Thermal Ceramics Ltd. He is also on the board of several major companies and a leading technical university.

Philips's story, post the mid-1980s, is particularly instructive. Founded in 1891, Philips became the largest consumer electronics company in the world after World War II. In the early 1980s, however, Matsushita overtook Philips and became the largest firm in the industry. Philips's saga of struggle for viability and growth over the last three decades is a result of its inability to cope with the challenges of globalization. Going international contributed to its dominance in consumer electronics, but paradoxically also became its Achilles heel.

One critical incident tipped Philips from dominance to an also-ran. Globalization was the backdrop to and the not-so-invisible hand in this story. Philips's national organizations (NOs)—rather than its global product-line divisions (PDs)—had become its powerhouses in terms of strategy. The best managers sought NO assignments to the glamorous capitals of the world, away from their rainy, small, placid, company town of Eindhoven, where the PD's top managers and Philips's headquarters were located. NO managers had direct links with the board of directors. They had command and control over the entire value chains in their countries, from R&D to sales and service. Their autonomous decision making and responsiveness to local markets propelled Philips to dominance in its industry. But this autonomy led to the critical incident that made Matsushita #1.

The head of North American Philips made a fateful decision relating to videocassette recorders (VCRs) that flipped the positions of Philips and Matsushita. Soon after they were developed, VCRs became, by far, the biggest revenue and the biggest profit-generating product in the consumer electronics industry. At the beginning of the product life cycle, competition among VCR manufacturers was intense. There were multiple incompatible formats jockeying for dominance. It was critically important that the format employed by a manufacturer become the biggest seller because the choice of format by videotape content producers, the sharing of videotapes among owners of VCRs, and other benefits of owning a VCR with a particular format were dependent on the installed base—the number of units with the same format in use. Manufacturers who succeeded in gaining market share would be assured of a further boost to sales as it benefited customers to own VCRs with a format many others were using. The battle of formats to become the standard was a matter of life or death.

There were four formats that relate to this story. In order of reputed quality, these four were developed, respectively, by (1) Philips; (2) Sony (Betamax); (3) JVC, a small Japanese company (VHS); and (4) Matsushita.

The Japanese Ministry of International Trade and Industry (MITI), in a surprise move, mandated that Matsushita adopt JVC's VHS standard. With Matsushita's huge domestic distribution network, sales of VHS format VCRs soared in Japan.

And then the unthinkable happened. In the midst of this global competitive struggle, where manufacturers fought to ensure that their format became the standard, the head of North American Philips chose to adopt Matsushita's VHS format rather than Philips's own format. Remember, North America then was and still is the largest market in the world for consumer electronics. And the rest, as they say, is history.

Think of the tangled causes and the irony! MITI forces Matsushita to adopt the VHS format. North American Philips chooses to sell its archrival's format rather than its own, catapulting Matsushita to #1. Betamax lost the battle with VHS, despite being of higher quality, because VHS was better suited to movie-length content. As it turned out, the viewing of "adult" movies was the primary reason why consumers bought VCRs.

The VHS-standard VCR was the reason for Matsushita's overtaking Philips and becoming the leading company in consumer electronics. But, a few years later, Matsushita fell victim to the Winner's Curse, and, curiously, the travails of Matsushita over the last decade can also be attributed to the VCR. At one time, more than 30% of Matsushita's revenues and 40% of its profits were generated by VCR sales! The sales of VCRs were arguably the main reason for Matsushita's ascendance. When DVDs took over from videotapes, Matsushita's fortunes plummeted along with its VCR sales, as they had nothing to replace their once-dominant product that they had come to rely on unquestioningly. Matsushita, today, is still struggling to recover.

Classic strategy development, with its linear problem-solving approach and touching faith in the possibility of predicting the future, could not possibly have dealt with the unprecedented, unpredictable, complex, and idiosyncratic developments that transpired throughout this story. Perhaps if Philips and Matsushita had employed the BoH approach to strategy development, they may have coped better with the globe-spanning complexities and uncertainties they faced. Most interestingly, after decades of struggling, Philips is currently adopting strategies similar to the BoH Strategy presented in Figure 1.2.

Coming back to our discussion of globalization, even companies operating in countries with large markets, like the United States, ultimately have to think and act globally when they—like McDonalds, Walmart, and Harley Davidson—have saturated the American market. But there is yet another and even more powerful reason why globalization is going to intensify in the foreseeable future. The economic center of gravity is shifting toward the East. The famed Goldman Sachs paper (Wilson and Purushothaman 2003) identifying the BRIC (Brazil, Russia, India, and China) economies as the giants of the future alerted us to this elemental shift.

Managers, who seek counsel from the writings and sayings of Niccolò Machiavelli, Sun Tzu, Yogi Berra, and Edward Murphy, might instead want to draw upon the wisdom of notorious bank robber Willie Sutton when considering the BRIC countries. When asked why he robbed banks, Sutton is reputed to have responded "because that's where the money is." This has led to Sutton's law, which is formally recognized in medical education, that "when diagnosing, one should first consider the obvious." It is obvious that the emerging economies, the BRICS (BRIC countries plus South Africa), are where the action is, where there is growth to be had and profits to be made. The BRICS have themselves recognized this and the leadership of these countries have started meeting regularly, with the intent of making their economies even more significant players on the global stage.

There are significant reasons why globalization is critical from the perspective of developing and implementing BoH Strategies:

1. **Pervasiveness:** Globalization affects all businesses. Even a company with no global aspirations has to take into account global forces, developments, and opportunities to remain viable in its geographically constrained market. Remember, global sourcing, global competition, global standards, global quality expectations, global partnerships, and global financing are important, even if the markets and customers are local.
2. **Intensification:** Per Sutton's law, globalization is inevitably going to intensify, facilitated by factors such as ongoing advances in communication and travel capabilities and the rise of BRICS.
3. **Complexity:** Globalization brings with it impenetrable complexity and incomprehensible uncertainty. Global planning is immensely difficult owing to enormous, complex differences across countries' cultures, ideologies, mores, and political, legal, educational, and

commercial institutions. And today's volatile and ever-changing geopolitical situations—in the Mideast, South Asia, and Eastern Europe, for example—leave us with virtually impenetrable uncertainty. The election of Donald Trump to the presidency of the United States greatly exacerbates the already unmanageable uncertainty.

INNOVATION

Innovation is an *imperative* for organizations. It is essential in order to respond to or evade competitive developments, anticipate customer expectations, cope with change, support growth, and enable sustainability. The maturity of markets in developed countries and the relatively slow growth rate in these markets place a premium on innovation by firms as an accelerant. In the emerging economies, innovation is also what enables firms to catch up with or leapfrog firms in the developed economies.

At the macroeconomic level, innovation is also an imperative. Economic growth results from two possibilities (Rosenberg 2004): increasing inputs or innovation. Unfortunately, increasing inputs, such as foreign direct investment, or investment from within the economy by firms, individuals, and the government may not always be a possibility. So, innovation, the second route to growth of a national economy, becomes very important. Innovation can come in the form of process or infrastructure efficiencies (by way of internal initiatives or governmental regulations) or in the form of differentiated or improved outputs. Champagne from France, single malts from Scotland, coffee from Colombia, chemicals from Germany, solar panels from China, advanced weaponry from the United States, and satellite launches from India, all boost their countries' economies through innovations in technology or marketing.

The recent Mars mission launched by India—called "Mangalyaan"—is a striking example of a country's innovative output. Because of its scientists' and engineers' innovations, as Prime Minister Narendra Modi pointed out, the Indian Mars mission cost the government less than it cost Hollywood to produce the space movie *Gravity*; in fact, 30% less, because of innovation. These innovations and capabilities, displayed in dramatic fashion by Mangalyaan, position India as a powerful competitor in the space launch business.

Innovative firms have a major impact on the national economy. The prestigious automobile firms in Germany—Audi (Volkswagen), BMW, Mercedes Benz, and Porsche—all contribute significantly through innovation to the growth of Germany's economy and, thus, make the country more competitive globally. President Obama's efforts to make the United States more competitive include boosting innovation in renewable energy.

In order to understand the implications of the mega-force of innovation on Wicked Strategies, it is necessary to recognize two kinds of innovation. The first is "sustaining innovation" and the second is "disruptive innovation" (Christensen 2013). Sustaining innovation enhances existing business models and responds to the needs of existing markets. Disruptive innovation creates new business models and, possibly, new markets.

While all innovation does not give rise to added complexity and uncertainty, disruptive innovations in technology and in business models pose a major strategic challenge (Christensen and Overdorf 2000). The trajectory and consequences of disruptive innovations are difficult if not impossible to predict, giving rise to uncertainty and ambiguity, which are better handled by BoH Strategies.

SHARED VALUE

Shared value is an *important* approach to managerial decision making that is gaining traction. It views Milton Friedman's (1970) singular focus on shareholder value as leading managers to make decisions that may be suboptimal and perhaps even damaging to the bottom line. Economic value and societal good are believed to be synergistic (Porter and Kramer 2011). Engaging a variety of stakeholders—employees, the community, customers, government, creditors, social activists—and responding thoughtfully to their perceived interests are expected to enhance economic and shareholder value, possibly in the short run and certainly in the long run. As a consequence, there is a growing recognition that shared value can and should inform and inspire the *raison d'être* of firms.

Shared value requires innovative planning processes. Traditional strategic planning processes (Camillus 1986) adopt a highly structured, linear approach to formulating strategy. Embracing a shared-value approach, which recognizes the different aspirations and priorities of multiple stakeholders, demands more complex and inclusive planning processes. These processes are the ones that are needed to formulate BoH Strategies.

Even if one does not subscribe to Porter and Kramer's (2011) assertion that economic value and social value are synergistic, with each enhancing the other, it is increasingly necessary to recognize other stakeholders because of their increased activism and potentially adverse impact. BoH Strategies are needed to generate shared value or to meaningfully respond to multiple stakeholders.

The growing power exerted by stakeholders other than shareholders is dramatically evident. For instance, Tata, India's most significant and revered (and we do mean "revered") company, had to abandon a new $300 million automobile assembly plant—for its breakthrough Nano model—built in an Indian State ruled by a *Communist* government because of the protest of farmers, egged on by *capitalist* politicians, whose land had been acquired to build the plant; a plant that offered a golden promise of jobs and economic development. Was this a challenge for which Tata should have developed and employed BoH Strategies?

In the United States, activist organizations, including college endowments and pension funds, which have substantial investments in companies, are pressuring companies to invest in green technologies and renewable energy, disinvest in Israel, support gay rights, invest domestically, and shun genetically modified organisms (GMOs). Socially conscious mutual funds attract investors even if their returns are below the returns of equity index funds. Communities do not allow big box stores to locate in their towns. Religious leaders pressure companies to adhere to principles and practices espoused by their religions. And the list of stakeholders who have the intent and possibly the power to shape the destiny of organizations goes on.

The Porter and Kramer (2011) argument that strategies based on shared value will increase profits should be enough to motivate firms to embrace this approach to strategy making. But, even if managers do not buy in to the Porter and Kramer (2011) proposition, or if they are unclear about how to incorporate shared value in their strategy making, they may not have any option but to recognize the importance of and respond to powerful stakeholders whose priorities are likely to be different from those of the majority of their shareholders.

SUFFICIENCY OF CONSIDERING THREE FORCES

Having looked at the indisputable relevance and substantial impact that globalization, innovation, and shared value have on firm strategy, the question raised at the beginning of this chapter remains: "Why do we focus on and limit ourselves to these three mega-forces?" Our discussion revolves around them for three main reasons:

1. **Importance:** As we hope the preceding discussion has demonstrated, each of these three mega-forces has a major impact on business strategies.
2. **Perspective:** Globalization, innovation, and shared value become major sources of wicked problems when they are entangled, compounding both complexity and uncertainty. (The interactions of these three mega-forces are the focus of the next three chapters.)
3. **Simplicity:** These three mega-forces encompass all the critical issues, subsuming the factors that are suggested by popular frameworks for guiding environmental analysis. Defining and evaluating additional forces will not illuminate the process of developing Wicked Strategies. And mention of additional forces would increase the complexity of the discussion immensely, with miniscule added value.

To demonstrate that the factors that are included in standard models of the environment are, in fact, components of these three mega-forces, let us consider the elements of the standard PEST model—political, economic, social, and technological factors. While the PEST (or STEP) model is widely accepted and powerful, and needs no explanation, we have found it necessary to add three elements, at the minimum, to this model—namely, (i) regulatory, (ii) ecological, and (iii) demographic factors. *Regulation*, while connected to the political context, has a distinct and direct impact on industries, transforming them in fundamental ways. In the United States, health, energy, communications, and financial services are examples of industries that are powerfully affected by regulation. *Ecological* factors have taken on profound importance for businesses because of social pressure, regulation, and the specter of climate change. And *demographic* characteristics—population, age distribution, income distribution—of countries substantially determine their ecological and economic challenges. Adding these three important factors to the PEST model creates a more comprehensive (REDPEST) model.

Table 2.1 presents a perspective on how the comprehensive list of REDPEST factors find expression in the mega-forces of globalization, innovation, and shared value. The greater number of factors in the REDPEST model is not necessarily helpful or needed to enrich our analysis because, as Table 2.1 suggests, the three mega-forces incorporate the effects of the seven factors.

TABLE 2.1

Illustrative Linkages between the "REDPEST" Factors and the Mega-Forces

REDPEST Factors \ Mega-Forces	Globalization	Innovation	Shared Value
Regulatory	Tariffs and trade agreements	Limitations (e.g., stem cell research; GMOs)	CSR mandates (e.g., in India; 2% of PAT)
Ecological	Carbon footprints	Sustainable technologies/ products	Community health
Demographic	Population and markets	Size of low-income population/market	Diversity/homogeneity of the population
Political	Alliances and ideological differences	Government-supported research	Form of government (dictatorship; liberal democracy)
Economic	Buying power of various countries	Resource availability	Income distribution
Social	Cultural mores of nations	Conceptions of success; individual freedom	Priorities accorded to various stakeholders
Technological	Offshoring	Focus and source of innovation	New value propositions and business models

The regulatory, ecological, and demographic (RED) factors that were added to the PEST model illustrate an important characteristic of the business environment that aggravates the challenge. The factors interact. It is difficult, if not impossible, to keep track of the 21 interactions between the seven factors in the REDPEST model. Our focus on the three mega-forces fortunately simplifies the task of understanding the interactions enormously, without entirely losing the advantages of the granularity of the REDPEST model, because, as Table 2.1 suggests, the mega-forces incorporate the effects of the seven factors.

INTERACTIONS OF THE THREE FORCES

While the individual impacts of *globalization, innovation,* and *shared value* are undoubtedly significant, it is the interaction and entanglement between these mega-forces that create both the challenge and the opportunity that demand BoH Strategies. These interactions and entanglements are illustrated in Figure 2.1 and described below.

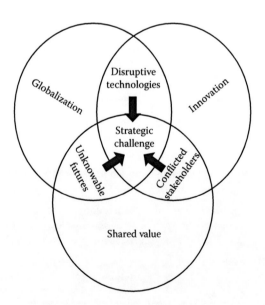

FIGURE 2.1
The interactions of the mega-forces.

THE INTERACTION OF GLOBALIZATION
AND INNOVATION

Globalization and *innovation* interact in significant and complex ways that result in *disruptive technologies*. Enormous emerging economies in difficult-to-access markets require entirely new business models that encompass redesigned products, lower price points, and novel distribution channels. Technological innovation makes large low-income populations increasingly possible to serve, and the recent dramatic shift in corporate R&D investments from local to emerging economies reflects this capability.

The data indicating the shift of corporate R&D investment to the BRIC economies is dramatic and compelling. A 2006 study by Yves Doz and several others found that firms now have 50% more research sites in foreign countries than their own countries of origin (Doz et al. 2006). At the time of the study, the investments of U.S. firms in R&D in China and India were on the verge of exceeding the firms' investments in Western Europe.

This pronounced shift, to China and India in particular, is readily understood through the lens of traditional strategic analysis. Qualified and competent researchers are available in these countries at a fraction of the cost that would be incurred in the Triad (EU, Japan, and the United States). And there is a not so obvious but consequential reason for the shift—proximity to the vast and vastly different markets in emerging economies facilitates researchers' ability to understand and respond to those markets.

According to Clayton Christensen, the guru of disruptive innovation, combining the research, product development, manufacturing, and marketing functions of a Triad-based firm with millions of potential customers with relatively low income and different needs in the crucible of a BRIC location can be expected to result in spontaneous combustion and the emergence of disruptive technologies, products, and, importantly, resulting new business models (Hart and Christensen 2002).

The economic motivation, of course, is the $5 trillion (2005 international dollars) to $13 trillion (the latter estimate employing 2004 Purchasing Power Parity) of untapped purchasing power of the four billion people at the base of the pyramid (Hammond et al. 2007; Prahalad 2006). But it is impractical and unreasonable to think of tapping into this immense reservoir of potential revenue and profits with the products and services offered to customers at the apex of the pyramid. Hence, price-point pressures trigger disruptive technologies and products. Christensen (2013) points out that classic strategies are equipped to handle sustaining innovation and are inherently ill-suited to respond effectively to disruptive innovation. BoH Strategies, on the other hand, can be designed to create value from disruptive innovation. We will discuss this in more depth in Chapter 4.

THE INTERACTION OF INNOVATION AND SHARED VALUE

The interaction of the imperative of *innovation* and the important commitment to *shared value* generates *conflicted stakeholders*. This is because innovation can enable and even require changes in the business model. New business models will generate value in novel ways, changing the relevance and importance of existing stakeholders and requiring a fresh and different understanding of how the value will be shared. Conflict between stakeholders can reasonably be expected to happen as a result. We discussed previously that the interaction of innovation and globalization can create disruptive technologies that initially appear to threaten a firm's viability. If embraced as the basis for innovative business models, however, disruptive technologies can prove to be a powerful source of competitive advantage and economic value. Conflicted stakeholders, if motivated and enabled to co-create value, can similarly serve as a source of competitive advantage and economic value.

THE INTERACTION OF SHARED VALUE AND GLOBALIZATION

The interaction of *globalization* and *shared value* entangles the power and priorities of multiple and diverse stakeholders interacting with the complexities and uncertainties of globalization. Consider how difficult it is for anyone to predict what will happen next year in the Middle East or in the United States. The acute uncertainty that accompanies globalization and the extreme complexity resulting from diverse stakeholders' perspectives and priorities are a recipe for decision paralysis. This interaction creates unpredictable, indeed *unknowable futures.*

This third interaction between the mega-forces appears to be very different from the other two interactions—*conflicted stakeholders* and *disruptive technologies*. The differences between *conflicted stakeholders* can spark creativity. Well managed co-creation offers the promise of encouraging interaction between important but conflicted stakeholders. Such interaction can be the path to identifying, acquiring, or developing needed competencies to develop new products and enter new markets, which generate economic value that benefits and reconciles conflicted stakeholders. *Disruptive technologies* present more of a challenge, but still offer the potential for adding value. Traditional management processes and decision analyses are inclined to favor sustaining innovation over disruptive innovation, opening the firm to the possibility of staying with a business model that could become obsolete because of disruptive technologies and business models embraced and introduced by competitors. But, if embraced and thoughtfully managed, disruptive technologies can boost competitive advantage and economic sustainability.

With regard to *unknowable futures*, unlike the situation with *conflicted stakeholders* and *disruptive technologies*, the potential for adding value is not readily apparent. *Unknowable futures* mean that traditional strategic planning processes are rendered ineffective. Widely used strategic planning techniques such as Five-Forces Analysis and Value-Chain Analysis, as conventionally employed, work best in contexts where the future can be understood and where predictions are not meaningless.

Strategic planning must transform as it faces a future that is unknowable.

Unknowable futures, while creating a challenge, also offer an opportunity. The chaotic ambiguity that characterizes *unknowable futures* can be viewed as resistant to analysis and a dead end or be grasped as an opportunity to fashion a desired future out of the ambiguous, chaotic, and hence flexible context. In order to craft BoH Strategies that create economic value in the context of *unknowable* futures, the strategic planning processes and techniques employed need to shift from prediction to a focus on transforming the organization and its environment. It is interesting and reassuring to note that both innovative business models and co-creation of value gain added significance in the context of *unknowable futures.*

REFERENCES

Camillus, J.C. (1986). *Strategic Planning and Management Control: Systems for Survival and Success*. Lexington, MA: Lexington Books.

Christensen, C. (2013). *The Innovator's Dilemma: When New Technologies Cause Great Firms to Fail*. Boston, MA: Harvard Business Review Press.

Christensen, C.M., and Overdorf, M. (2000). Meeting the challenge of disruptive change. *Harvard Business Review*, 78(2): 66–77.

Doz, Y., Wilson, K., Veldhoen, S., Goldbrunner, T., and Altman, G. (2006). *Innovation: Is Global the Way Forward*. INSEAD, Fontainebleau.

Friedman, M. (1970, September 13). The social responsibility of business is to increase its profits. *The New York Times Magazine*, The New York Times Company.

Hammond, A.L., Kramer, W.J., Katz, R.S., Tran, J.T., and Walker, C. (2007). *The Next 4 Billion: Market Size and Business Strategy at the Base of the Pyramid*. World Resources Institute International Finance Corporation, p. 3.

Hart, S.L., and Christensen, C.M. (2002). The great leap: Driving innovation from the base of the pyramid. *Sloan Management Review*, 44(1): 51–56.

Porter, M.E., and Kramer, M.R. (2011). Creating shared value. *Harvard Business Review*, 89(1/2): 62–77.

Prahalad, C.K. (2006). *The fortune at the bottom of the pyramid: Eradicating poverty through profits*. Upper Saddle River, NJ: Wharton School Publishing.

Rosenberg, N. Innovation and economic growth. OECD 2004. Available at http://www.oecd.org/cfe/tourism/34267902.pdf

Wilson, D., and Purushothaman, R. (2003). Dreaming with BRICs: The path to 2050 (Global Economics Paper No. 99). Goldman, Sachs & Company.

3

Responding to the Strategic Challenge

The task of Business of Humanity (BoH) Strategies is to overcome the three challenges created by the interaction of the mega-forces, namely, disruptive technology, conflicted stakeholders, and unknowable futures. BoH Strategies are designed to convert these challenges into opportunities for enhanced sustainability on all three dimensions—economic, environmental, and social.

The BoH Strategies that we propose are intended to address the strategic challenge in these ways:

1. Crafting innovative business models that embrace and exploit disruptive innovation.
2. Engaging in co-creation of value with conflicted stakeholders to disarm them.
3. Envisioning and enabling a desired and sustainable future that replaces the chaotic ambiguity of an unknowable future.

Figure 3.1 captures the intent of BoH Strategies.

FROM DISRUPTIVE TECHNOLOGIES TO INNOVATIVE BUSINESS MODELS

In Chapter 2, we discussed how globalization and innovation interact to create disruptive technologies that need to be harnessed to create value. Disruptive technologies that are disregarded can render organizations' business models obsolete.

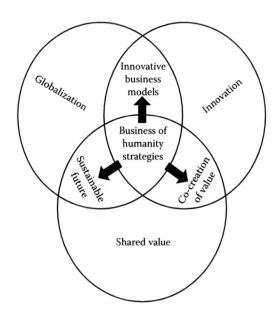

FIGURE 3.1
The intent of BoH Strategies.

To understand how to exploit disruptive technologies, let us begin with the powerful example of how one professor realized his vision of bringing telephonic communication to rural India. The story* revolves around Professor Ashok Jhunjhunwala of the Indian Institute of Technology, Madras, one of India's foremost engineering schools. Jhunjhunwala's accomplishments have been internationally recognized and honored by the Government of India with the prestigious national Padma Bhushan award. An expert in communications technology, Jhunjhunwala motivated a team of his computer and electrical engineering colleagues to join him in an ambitious venture to bring modern communication to rural India. At the time, India had a population of one billion people, but a miniscule 10 million telephones, almost all of which were city-based landlines. Jhunjhunwala's visionary goal was to increase coverage to 100 million telephone lines, which would also be made accessible to the rural population.

His organization—which was incorporated with the name TeNet—found that it took more than advances in technology, no matter how technically brilliant, to be successful. They found that even dramatically lowering price points, while it did make the communication products and services more accessible, was also not enough. The lower costs did

---

* See: http://www.katz.pitt.edu/boh/case-studies/bttm-pyramid.php

stimulate technological breakthroughs, but Jhunjhunwala found that the team's accomplishments and progress toward its goal were minimal until a *eureka* moment occurred.

Becoming One with the Customer

Jhunjhunwala, pondering the lack of progress, suddenly came to the conclusion that their singular focus on communications technology was holding them back. He realized that for the impoverished rural population, communications technology was of little interest in comparison to larger concerns. What rural Indians wanted and needed, he said, was "health, education, and livelihood." This might seem to have been obvious from the start, but it is typical of companies to blithely assume that what they have to offer is what new customers want, especially if another higher-income group was receptive to it.

Jhunjhunwala's epiphany on the psychology of low-income customers led to several developments. His company fashioned innovative value propositions based on harnessing communications technology to provide, for example, programmed primary and secondary education, basic telemedicine services, and new jobs for managers of local kiosks that would offer the telecommunication services. As TeNet expanded its operations, many outside businesses were positively affected. Companies were formed, investors were attracted, rural India benefited, and Jhunjhunwala ultimately received one of the highest honors from the Indian government for his technological, economic, and social contributions.

Responding to the opportunities created by Jhunjhunwala's efforts, other companies began developing complementary technologies and products. For instance, efficient algorithms were developed that compressed voice recordings into packets of data that were much smaller than what had previously been possible. This novel disruptive technology helped address the communication needs of the largely illiterate rural population. A form of e-mail communication between individuals who could not read and write was made possible by this development, which, using basic, slow, and cheap computers, efficiently recorded and communicated voice rather than written messages. This compression technology was extended to video, and one of Jhunjhunwala's companies was able to develop and build DSL routers that could, amazingly, send 11 movies simultaneously through the ancient, degraded copper wire that served the few telephones that existed in rural India.

It was abundantly evident, as Jhunjhunwala discovered, that disruptive technologies by themselves were not enough to bring about change, or to provide economic or social benefit. The value of disruptive technologies lies instead in their potential to offer new value propositions, enable supporting processes, and generate new profit models. Furthermore, to create effective value propositions, the needs of customers who are very different from the customers that companies are accustomed to serving have to be insightfully divined and addressed. This is why local research and direct R&D interaction with the target customer-base generate value far beyond cost savings.

There is a critically important lesson to be derived from Jhunjhunwala's experience at TeNet. To develop successful BoH Strategies, we must walk in the customers' shoes. The kiosk operators that TeNet set up in rural India were both part of TeNet's operations and were staffed by the archetypical customer that TeNet was seeking to serve. The kiosk operators provided an understanding of customer needs that no outsider could possibly offer. They functioned as distributors and providers of TeNet's services and products; they served as test markets and sources of insights about TeNet's customers' needs; and they were the beneficiaries of TeNet's strategic commitment to providing livelihood opportunities.

Integrating Social Responsibility into the Business Model

Becoming one with the customer at the base of the pyramid demands that social responsibility be a guide and a driver of the business model. Without a deep and fundamental commitment to the constructive role that business can play in benefiting humanity, it is tempting to dismiss the challenges of serving the bottom of the pyramid because they appear too daunting.

There is, however, substantial evidence that meeting basic needs of low-income segments can be rewarding in economic terms in addition to being psychologically rewarding. A powerful example of the symbiosis of economic, environmental, and social sustainability is that of Arvind Limited, a global textile manufacturer based in India, which reduced the incidence of cotton farmers committing suicide—a tragedy that is all too common in India.

The farmers who were most at risk worked on nonirrigated land, dependent on monsoon rains. They borrowed money to pay for expensive genetically modified, pest-resistant seeds, and for fertilizers and pesticides.

If successive monsoon rains failed, the farmers were likely to be unable to pay back the money they borrowed and their land would be forfeited to the moneylenders. With no ability to make an alternative living, farmers and their families often chose suicide to slow death by starvation.

Arvind, a public limited company, originally focused on manufacturing textiles, has the fourth generation of the founding family as its CEO. Arvind's reputation for leading management practices has made it the focus of several case studies that are discussed in leading management schools in India. Every generation of the family has been recognized for its commitment to social responsibility, a commitment that has seen a progression from personal giving, to corporate giving, to professionally managed philanthropic foundations funded by the company, to the current generation's innovative and surprisingly profitable incorporation of social responsibility into business models implemented by Arvind.

Motivated by their concern for the farmers, and by the company's and its major customers' (including Walmart and Patagonia) avowed commitment to environmental and social sustainability, Arvind's CEO and senior executives brainstormed a possible response to the farmers' precarious circumstances. The answer they came up with was to teach and enable the farmers to engage in organic cotton farming.

Organic farming got farmers out of the clutches of moneylenders because organic farming is fertilizer and pesticide free. Even the cost of cotton seed was significantly reduced because the expensive genetically modified seeds were not employed. Arvind's project took enormous commitment, a great deal of courage, some creativity, and rigorous planning and analysis. A detailed business plan was developed. The company recruited 130 agronomists and embedded them in the villages to offer scientific counsel tailored to each farmer's small plot of land. Arvind also assumed control of the value chain surrounding the farmers' activities, eliminating avaricious and exploitative middlemen, and fundamentally changed traditional farming practices.

But in order to achieve this splendid success, Arvind executives truly had to become one with the farmers. The son of the CEO, a Yale-educated, immensely advantaged fifth-generation scion of the company's founding family, lived in the villages. The head of Arvind's agribusiness also lived in the villages. Through these experiences, they experienced the hardships and challenges faced by the local population. It is impossible for anyone unfamiliar with rural India to understand the resolve it takes for someone used to middle-class levels of comfort, let alone a person of privilege, to endure the conditions of living in a poor village. Yet, Arvind undertook these efforts to understand the local population even further.

Beyond helping farmers to protect their lives and livelihoods, Arvind also protected the environment, engaged many diverse stakeholders, enhanced its competitive position and standing with major international customers, and added significantly to its coffers. The moral is that Arvind's leadership became one with the farmers and the resulting tale of disruption, increased economic value added, and social benefit is illuminating and inspirational.*

Arvind has committed to and replicated the practice of incorporating social responsibility into business models to create disruptions and innovatively enhance both economic value and social benefit. One of these innovative businesses is real estate development and housing for slum dwellers. Another involves teaching tailoring skills and providing safe and comfortable housing to economically and socially disadvantaged young women, to ease Arvind's entry into the Indian garmenting industry, which is challenged by highly restrictive labor laws.

Innovating across the Entire Value Chain

In each of the above examples, there is a related point of significance, and it is that disruptive technologies may not be necessary to create innovative business models. Other elements of the value chain can access untapped economic value and strengthen competitive advantage in new markets. These elements could be as simple as new packaging or as complex as creating a distribution system that accesses remote customers and shares value with the customers and other stakeholders.

Innovation across the entire value chain can be promoted by engaging customers and other stakeholders in the creation of value. Richard Normann and Rafael Ramirez (1993) describe a constellation of value created by "suppliers, business partners, allies, customers—work[ing] together to co-produce value."

Let us consider the example of CavinKare, which began as a personal care start-up called Chik India, and is now a highly successful fast-moving consumer goods (FMCG) manufacturer. The company started with an initial investment of less than $300 in 1983 and grew to a $250 million

---

* Arvind's story is told by its chairman and CEO, Sanjay Lalbhai, in a video that is available at https://www.youtube.com/watch?v=VPC8Kn_XaSA.

company by 2014. It got its start with the now-ubiquitous, single-use sachets of shampoo. These sachets were affordable for the hundreds of millions of customers earning less than $2 a day, who could not afford to buy a whole bottle of shampoo. A simple innovation in packaging led to a resounding success. CavinKare also offers another lesson. Its distribution system was and continues to be unique, and in 2014, CavinKare sought to give its sales and profits a boost by making major investments in its distribution system, embracing literally millions of little shops.

Distribution is of enormous importance in emerging economies. Companies have learned not to allow anyone or anything to come between them and the customer. Consequently, they have built distribution systems that tie them closely to the customer. Philips, for instance, in line with its new focus on health and lifestyle, built a distribution system in India that employs thousands of small shopkeepers and distributors who are allocated small territories. The employees are also potential customers and thus provide valuable insights to Philips. The company's distribution system, then, does more than serve the supply chain well; it is also carefully aligned with customer interests, and is a key source of competitive advantage for Philips' new low-cost personal care and kitchen products that are designed and built in India.

Another evocative example of the potential in innovating across the entire value chain is that of Vodafone CZ. Operating in the Czech Republic with two larger competitors, Vodafone CZ set a remarkable goal of "transforming the telecommunications industry."* This might seem like hubris, for a relatively small company, operating in a small country, with no in-country R&D capability in telecommunications. But the ambitious goal resulted in Vodafone CZ thinking creatively along the elements of the value chain that it did control. To meet this goal, Vodafone CZ adopted a value proposition that stated simply, "Break all the rules for the customer." Following the adoption of this value proposition, Vodafone CZ introduced practices that transformed the telecommunications business. Customer representatives were empowered to do anything that was required, even break company policies, to remedy problems encountered by customers. There was no limit to the monetary compensation that a representative could offer a customer who had encountered a problem caused by Vodafone CZ. Customers were not required to sign contracts and could, if they wished, shift to another carrier without penalty. Existing customers

---

* See: http://www.katz.pitt.edu/boh/case-studies/vodafone.php

were given all the incentives offered to new customers. Vodafone CZ publicly challenged its competitors to adopt its approach to serving the customer. And Vodafone CZ became, for its size, the most profitable tele-communications company in the world.

Transforming the Management System

A related lesson about BoH Strategies can be drawn from GE's experience. Jeffrey Immelt, GE's CEO, along with two academics, Vijay Govindarajan and Chris Trimble, recently wrote a fascinating and immensely instructive article (Immelt et al. 2009) that provided a radical perspective on competitive strategy and the management system needed to support it. They described the experience of GE's medical systems business in developing major breakthroughs in the cost and functionality of electrocardiogram (ECG) and ultrasound machines when engineers and scientists were charged with developing machines that could be profitably sold at a price point suitable for the low-income segments of the Chinese and Indian populations. The ECG machines developed in India were battery powered, because the electric grid did not reach into every village. The machines had to be small enough to fit on the back of a bicycle, which was often the only viable mode of transportation in rural areas. The machines had to be simple enough to be operated with little training, and they had to have telemedicine capabilities that would enable doctors in distant locations to review the data and assess the patient's condition, and, of course, the machines had to be made at a fraction of the cost of the existing products that had been developed to serve the higher-income populations of more developed countries.

The fact that GE accomplished these goals reinforces the lesson that Professor Jhunjhunwala learned. The technology must be driven by customer and contextual needs. Management should not assume that, if the technology is improved and achieves engineering-oriented, higher-performance goals, competitive advantage and business success will inevitably follow. Marginal improvements in existing technologies are unlikely to give rise to breakthrough or disruptive products and business models. Disruptive technologies are ones that make existing technologies obsolete—like how e-mail made faxes obsolete. However, existing technologies that are newly brought to bear to serve the specific needs of a market can also create a competitive advantage—for example, replacing viewfinders on video cameras with LCD screens, or adding telemedicine capabilities to ECG machines.

The GE example, most importantly, also drives home an absolutely critical point. For an organization to support the development of disruptive technologies and related business models, it must modify or transform its management system.

GE has been at the forefront of management expertise since its investment of $4 million (around $45 million in today's dollars) in the early 1950s in a project to identify the key business areas in which companies must perform well in order to be sustainable. This management expertise displays itself in GE's reworking of its management system to support its new strategic emphasis on emerging economies and reverse innovation.* GE explicitly adopted the strategy of meeting basic needs—water, infrastructure, health—in emerging economies and the strategy of reverse innovation to gain competitive advantage in developed economies.

According to Immelt et al. (2009), GE revolutionized its management system in order to support the development of breakthrough technologies, which in turn enabled new business models that boosted economic value added. The intent was to develop technology and business model innovations in emerging economies to add value in those markets, and to also bring back these developments to the Triad economies for added competitive advantage and profits. To implement this strategy, GE changed its organization structure to bring new attention to the fast-growing emerging economies, modified its resource allocation and performance measurement practices to focus on disruptive rather than sustaining innovation, and reaffirmed its commitment to an entrepreneurial, risk-taking management style.

GE had traditionally viewed its financial performance by product line, focusing on the global revenues and profits of each product line. In this scheme, each of the emerging economies contributed a small amount, thus not meriting much management attention or resources. In order to support its strategy of reverse innovation based on a growing emphasis on emerging markets, GE reoriented its reports to focus on revenues and profits by country, which increased the significance of each country and emerging economy in the management's deliberations. Country managers gained significance in the organizational matrix, and consequently they were better able to direct resources, human and financial, from across the

---

* Reverse innovation is the process of developing products and technologies in emerging economies and then taking them to the developed economies. It is the reverse of the traditional process of "glocalization," which is taking products from developed economies and modifying them for sale in emerging economies.

company to support innovation in and for emerging economies. Even the measures of performance changed to support disruptive innovation.

For instance, Immelt expressed reservations about market share, which has traditionally been viewed as the most important and meaningful measure of competitive strength and strategic performance. The reason for his reservations, beyond the usual concerns about how to define the market and the changing nature of the market, were that the market share measure kept management's attention on the past. The larger the market share, the less inclined managers would be to accept and employ disruptive innovations. Instead, they would see sustaining innovations as the only logical way to go, in order to build on their existing strength. And as Christensen (2013) points out, traditional incremental economic analysis is strongly weighted in favor of sustaining innovation. The fixed costs and investments needed to create disruptive innovations will, justifiably, be entered into a decision matrix. In the case of sustaining innovations, such investments have often been already incurred and will also justifiably be treated as sunk costs and, as such, will be irrelevant to the analysis. However, managers conducting the analyses and committing to sustaining innovation will not be predisposed to recognize and anticipate the potential impact of disruptive technologies, which could very well abruptly end the cash flow projected from existing operations.

GE reworked its performance measurement, planning processes, and organizational structure—in other words, its management system—in order to implement its strategies, which substantially align with the BoH elements of global responsiveness, meeting the basic needs of low-income segments, and, in essence, seeking synergy between economic value and social benefit.

Stretch Goals, BHAGs, and Frugal Engineering

A fundamentally important, common thread running through the examples we have discussed involves the importance of big, hairy, audacious goals (BHAGs*) to drive and direct breakthroughs in technology and business models. BHAGs force individuals to break out of the limitations of linear thinking. In emerging economies, the pressing and basic needs and the limited purchasing power of the base of the pyramid force firms to explicitly articulate and commit to BHAGs.

---

* The term BHAGs—pronounced bee-hags—is the acronym derived from "Big Hairy Audacious Goals." It was coined by Collins and Porras (1996).

In the late 1990s and early 2000s, five benchmarking studies (four organized by the APQC, previously known as the American Productivity and Quality Center) and one sponsored by the Hong Kong Productivity Council researched the practices of 87 companies, including 22 "best-in-class" companies from across three continents. One of the most striking findings was that the best-in-class companies stressed the importance of stretch goals in improving performance significantly more than the companies who served as a control group.

BHAGs have been the motivator of many notable breakthrough innovations. The Tata Nano, the least expensive car in the world, is a stunning feat of engineering. Ratan Tata, then chairman of the Tata Group, was reputed to have set a goal of producing a $2000 car when he saw a family of four on a motorcycle involved in an accident. Motorcycles commonly carry entire families in countries with low per-capita income.

In emerging economies, the key BHAG that drives BoH Strategies is honoring the price point that would enable large, low-income segments to become paying customers. This is a corollary of becoming one with the customer. Frugal engineering—which is based on intensive value engineering at the design stage, and rigorously applied lean engineering in the manufacturing process, supported by intensive quality management—is needed to meet the price point challenge. The practice and processes of frugal engineering are key to the success of BoH Strategies and will be discussed in detail in Chapter 7.

There are two relatively novel approaches in management practice and thinking that can assist in implementing frugal engineering processes. First is the design thinking (Brown 2008) process, which seeks to build empathy with the client or customer and stimulate creative responses to the customer's needs. Second is the increasing power of analytics focused on big data (Chen et al. 2012). While analysis of big data cannot reveal cause–effect relationships, businesses now are able to derive powerful insights (Strong 2015) about the customer that can support innovative value propositions. The combination of design thinking and big-data insights can facilitate and enhance the practice of frugal engineering.

Compiling the Responses to Disruptive Technologies: Building Innovative Business Models

Perhaps the most important lesson that can be derived from looking at these examples is that what appears to be a dire threat to organizations can instead, with a ton of courage and a modicum of creativity, be alchemized into a source of competitive advantage and added economic value.

So far, in this chapter we have discussed some critical practices for handling and embracing disruptive technologies:

- Becoming one with the customer
- Accepting unique and unserved customer needs as a driver
- Innovating across the value chain
- Employing stretch goals
- Reworking the management system
- Employing design thinking and data analytics

We can readily translate these practices into guidelines that enable organizations to transmute disruption and chaos into cash flow:

1. Specifying extraordinarily ambitious goals—especially price point–related BHAGs
2. Relating empathetically to the customer
3. Integrating social responsibility into the business model
4. Exploring disruptive possibilities along the entire value chain
5. Implementing supportive organizational structures and planning and control systems—especially frugal engineering processes incorporating design thinking and data analytics

These responses together combine to create *innovative business models* that incorporate and exploit *disruptive technologies*.

We will elaborate further on the necessary characteristics and process of developing BoH Strategies, after first exploring the implications of *conflicted stakeholders* formed by the intersections of innovation and shared value, and *unknowable futures* formed by the intersection of shared value and globalization.

FROM CONFLICTED STAKEHOLDERS TO THE CO-CREATION OF VALUE

Co-creation is a consciously managed, carefully defined effort to promote value-creating innovation. This effort involves collaboration by a selected set of individuals who possess relevant knowledge or perspectives. Possibilities for such co-creation abound. In the context of developing BoH Strategies, co-creation can focus on the following:

- Elements of the business model, including an organization's value proposition, espoused goals, and performance-measurement standards
- Individual components of the generic value chain; both supporting activities, such as technology, and primary activities, such as marketing and distribution
- Partnerships with other firms (e.g., suppliers, customers, and alliance partners)
- Interactions between departments and functions within the firm
- The entire range of stakeholders who are significant in terms of their potential impact on the firm, or who are affected by the activities of the firm

We have already discussed the necessity of becoming one with the customer. Value co-creation with customers is common in efforts related to innovations in branding, product and service variations, and packaging. But, if the firm is committed to *shared value*, co-creation of value with other stakeholders, in addition to customers, is a must. The "value constellation" proposed by Normann and Ramirez (1993), which was mentioned earlier, describes engaging a range of stakeholders in "co-producing" value at various stages in the firm's value chain.

Most pertinent to our discussion is the notion of "innovating for shared value" (Pfitzer et al. 2013). In addition to unearthing opportunities for added economic value, effective engagement of stakeholders can enhance their sense of ownership and identification with the firm. Co-creation of value results in partnerships that have the possibility to offer a firm access to new resources and influence.

To understand and support co-creation of value, we will consider the following:

1. Earning and receiving value from co-creation
2. Managing stakeholders involved in the co-creation process
3. Building alliances to co-create value
4. Promoting co-creation of value within the firm
5. Forming an innovation ecosystem to accelerate co-creation of value

Earning and Receiving Value from Co-Creation

Building long-term partnerships and value is important to tapping the potential synergy between innovation and shared value. Consequently, the co-creation processes we need to design and employ are quite different from open-sourcing and crowd-sourcing processes. Open sourcing and crowd sourcing tend to be technology based and do work well in certain situations. Open sourcing works well when what is being sought is a plethora of ideas for something like a brand name, or small contributions from individuals.

The motivations of participants (Mehrpouya et al. 2013) in open-sourcing approaches are usually personal, are competitive with regard to other participants, and are not necessarily conducive to a long-term involvement. These motivations are not meaningful for the types of effective, stakeholder-serving innovation that is the purpose of the co-creation in which we are interested.

An example of co-creating an organization's value proposition was an exercise conducted by one of our client firms, whose growth and profits were stagnating. The situation became untenable when the firm's survival was threatened by unforeseen developments triggered by the explosion of global terrorism. The CEO engaged in intense planning sessions with a wide range of personnel in the firm, but was unable to arrive at a promising strategy to respond to the context. The firm's CEO came to the conclusion that fresh thinking was required. His solution was to have a detailed case study of the firm prepared, detailing its history, strategy, resources, and the changing context that it faced. The CEO, employing his own and his consultant's contacts, identified 15 executives from other companies who had a reputation for creative thinking. The C-suite executives from the firm, as well as several managers from various departments and functions, joined the 15 invited executives to discuss the case study, with the goal of identifying mission and strategic alternatives as well as possible significant operational enhancements. A host of credible ideas emerged during the discussion. These ideas were later vetted by the CEO and the consultant, and then reviewed by a large number of managers in the firm. As a result of this exercise in co-creation, the firm adopted a promising new value proposition and also decided to implement several of the operational enhancements that had come up during the discussion.

Managing Stakeholders

Inviting executives from other firms to participate in a co-creation process is an unusual practice, though it worked very well for this firm. Stakeholders such as suppliers, distributors, and customers are usually involved in co-creation exercises. When identifying stakeholders to participate in such exercises, it is useful to employ a classification scheme such as the matrix presented in Figure 3.2.

A number of different stakeholder-management approaches should be considered when making any strategic decision:

- Moving the powerful, adversarial stakeholders to a different relationship by providing incentives (arrow "1" in Figure 3.2)
- Offering opportunities for the powerful, adversarial stakeholders to be better and more positively engaged with the firm (arrow "1" in Figure 3.2)
- Strategically repositioning the firm (arrow "2" in Figure 3.2) to reduce the power of adversarial stakeholders
- Strategically repositioning the firm (arrow "3" in Figure 3.2) to make supportive stakeholders more significant

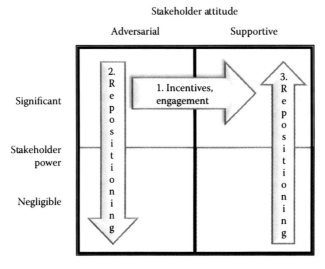

FIGURE 3.2
Stakeholder classification matrix.

- Creating opportunities for even better engagement with powerful supportive stakeholders, and repositioning the firm to enhance their power

This matrix offers relevance and value in the context of selecting a country to enter or deciding where to locate a plant. It can also help improve strategic decisions where the identity and potential role of stakeholders is not readily obvious.

Building Alliances

Alliances provide a boost to co-creation. They are particularly helpful when entering into or creating new markets. Alliances are a way of acquiring or accessing competencies that are needed to handle the disruptions likely to occur in emerging economies.

Enhanced value co-creation through alliances happened in the example we discussed previously in which Arvind taught destitute farmers how to grow organic cotton in nonirrigated lands that were at the mercy of the vagaries of the monsoon season. To increase the value added by its project with the farmers, Arvind negotiated help, in the form of a system of irrigation ponds, from long-term partners such as the Tata Group of companies and the Gujarat State Government. The government, of course, was motivated by the social and environmental benefits of the project. Tata's long-held core values found expression in its contribution to the economic well-being of the farmers.

In building alliances for co-creation, Rosabeth Moss Kanter (1994) offers some important guidelines. She urges firms to partner only with the finest organizations. Effective alliances can only be built on a foundation of trust. Having the finest partner with which to work motivates and creates such a climate.

Kanter (1994) identifies levels of interaction between partnering firms that suggest opportunities for co-creation. The levels suggested are (1) strategic, (2) tactical, (3) operational, (4) interpersonal, and (5) cultural.

The *strategic level* suggests co-creating new markets. An example is the catalog flower business that was created by Ruth Owades. Her start-up company, Calyx and Corolla, became a great success because of the alliances she carefully instituted and nurtured. Her experience in the catalog business gave her the credibility to pull together FedEx and pioneering flower growers to create an entirely new business model for selling flowers.

This strategic level of co-creation develops and affirms a common understanding of the new target market, which is shared by firms in the alliance.

The *tactical level* refers to the projects that the partners work on together to exploit their target market. Both the different and complementary competencies possessed by the partners sustain the projects. Differences could be a source of friction, but, in beneficial partnerships, they make the alliance more productive. The *operational level* focuses on the processes that need to be in place to support co-creation.

Communications technology could play an important role here. The *interpersonal level* addresses the relationships between individuals across the companies in the alliance. Often, a close relationship between champions of the project (*tactical level*) in each of the companies involved is what initially sustains the alliance and supports co-creation. Finally, the *cultural level* recognizes that differences in perspective and knowledge are what make co-creation happen; there has to be mutual respect for and acceptance of the partner's differences, not responses that cause alienation.

Opportunities for co-creation through partnerships can be identified by examining the firm's value chain. Co-creation fostered by connecting elements of the firm's value chain with those of suppliers or customers can be very productive. Alcoa has several excellent examples of such co-creation of value. The company partnered with Audi to create a breakthrough, space-age automobile frame. Alcoa also worked with Pittsburgh Brewing to create a unique, deep-drawn aluminum bottle, the novelty value of which greatly boosted the sales of the IC Light brand of beer, and, in Alcoa's recent and highly visible partnership with Ford, the companies worked together to replace steel with aluminum in the body of Ford's most important and profitable product, its F-150 pickup truck.

There are several possible connections across the value chain that can support co-creation. The design of components or products can be coordinated with the manufacturing or marketing functions of the customer. Logistics and distribution can be linked with the purchasing function of the customer.

Promoting Co-Creation within the Firm

Many companies co-create value internally through project teams with diverse membership. The famous Team Taurus, which created one of Ford Motor Company's most successful models, included engineers with a variety of expertise: designers, marketers, and accountants. It is quite common for firms to follow this "super-team" approach to commercialize new technologies. Lockheed Martin, for example, has its "Skunk Works"—an alias inspired by a moonshine factory of the same name in the comic strip Li'l Abner—which are "super-teams" located physically separately from the main operations of the firm, sometimes on a secret or confidential basis, to create major new and superior products. The Lockheed Skunk Works produced breakthrough aircraft such as the U-2 and the SR-71 Blackbird, which are considered to be some of the most advanced aircraft ever made. It is interesting to note that breakthroughs seem to require a great deal of autonomy and a separation from the main line of business. IBM, for instance, found that it had to develop the PC using a small team, kept away from its headquarters in Armonk, New York. IBM was very much a mainframe manufacturer in those days, and the PC would have found little support at headquarters.

Skunk works and super-teams function as collateral organizations, separate from the main organization. In effect, they are delinked from the inertia, or the directional vector, of the primary organization. BoH Strategies, because they are different in many ways from traditional strategies, may need to be fostered by such skunk works and collateral organizations. Skunk works and super-teams insulate entrepreneurial ventures and transformational innovations from the primary organization's emotional investment in and commitment to existing technologies and business models.

It is critically important to recognize that BoH Strategies would require such protection when first introduced in an existing firm with a history of employing traditional strategies.

Forming Innovation Ecosystems for Co-Creating Value

Another approach to innovation involves creating an ecosystem consisting of organizations that together create the conditions for stimulating and nurturing radical innovation. The highly interactive diversity of capabilities that drives innovation in skunk works also works for groups of companies. This engine of innovation has its roots in a framework that Michael Porter (2011) highlighted when he analyzed the competitive advantage of nations. He identified the importance of related and supplier industries, which are the foundation for the formation of industry clusters such as Hollywood and Bollywood for movies, Detroit and Stuttgart for automobiles, and Silicon Valley and Boston for information technology.

Clusters are a static concept that evolved into the notion of innovation ecosystems. Innovation ecosystems are a dynamic mix of small companies, large companies, start-ups, diverse services, venture capital, research institutions, universities, and supportive government policy. Innovation ecosystems are a hotbed of innovation and co-created value.

Innovation surges when connections are made between ecosystems in different industries. Examples include dramatic innovations resulting from connecting the automobile and information technology industries. Similarly, connections between IT and healthcare have boosted innovation at all stages of the health industry value chain from research to client records.

There is a related phenomenon, which is of particular importance to BoH Strategies. It has been observed that connecting ecosystems in the same industry, but which operate in different countries, stimulates innovation. This phenomenon is what companies like GE and Philips are relying on to support their goals of reverse innovation. BoH Strategies, which are inherently global in nature, need to exploit the benefits of connecting innovation ecosystems across markets and countries that possess different characteristics.

To take advantage of this phenomenon, global communities of knowledge can be set up to foster innovation. Royal Dutch Shell has for many years utilized virtual communities of knowledge to share information about best practices in planning across countries. Another example with which we have personal familiarity is the Low Voltage DC (LVDC) Forum of the Institute of Electrical and Electronics Engineers (IEEE), which brings together companies, universities, research institutions, consultants, and government bodies. The members of the LVDC Forum seek to establish standards, engage with government regulators, and jointly implement pilot projects. They share the knowledge emerging from the research conducted by individual members, which could support their common objectives and programs. These communities of knowledge are powerful catalysts and accelerators of innovation, as well as the commercialization of innovation.

Compiling the Responses to Conflicted Stakeholders: Co-Creating Value

Co-creation processes can convert conflicted stakeholders into engaged partners who promote creativity and innovation and enable BoH Strategies. Co-creation can be employed to develop novel value propositions, inspiring new visions and innovative business models that integrate economic value and social benefit. Differences between stakeholders become a desirable condition rather than a source of conflict. Co-creation thrives on differences, and these differences enable the innovations that support BoH Strategies. Of course, it is necessary to identify and manage the relevant stakeholders who, though different, need to be complementary in capabilities and mutually respectful.

Based on what we have discussed in this chapter, certain conditions are required in order for co-creation to occur:

1. Stakeholders need to be carefully managed, which is to say consciously identified, analyzed, and co-incentivized. The priorities of powerful, adversarial stakeholders need to be recognized and addressed.
2. Stakeholders, including suppliers, business partners, and customers, can be engaged in co-producing value at any or all stages in the value chain.
3. Alliances need to be built with firms that have complementary capabilities.
4. The structural context in which co-creation takes place within an organization is important. Separation from existing strategies and traditional mindsets may be necessary to enable and nurture BoH Strategies.
5. Innovation ecosystems provide a fertile field for growing BoH Strategies.

Co-creation, because it proactively engages conflicted stakeholders and increases the organization's capability to manage and potentially benefit from disruptive innovation, can offer powerful support to BoH Strategies. The third and final challenge—*unknowable futures*—is discussed below.

FROM UNKNOWABLE FUTURES TO FEED-FORWARD

The maelstrom of extreme complexity and uncertainty that exists at the intersection of the forces of globalization and the mandates of shared value make predicting the future an impossibility. Extreme complexity and uncertainty are also a breeding ground for problems that are classified as "wicked."

Horst W.J. Rittel and Melvin M. Webber, professors of design and urban planning at the University of California at Berkeley, first described wicked problems in a 1973 article in *Policy Sciences* magazine. They identified 10 characteristics of wicked problems. Over the years, these 10 criteria have been reduced (Camillus 2008) to five essential characteristics that make traditional problem-solving approaches impotent:

1. The perceived "problem" is unusual and substantially without precedent.
2. There are multiple, significant stakeholders with conflicting values and priorities who are affected by the perceived "problem" or responses to the "problem."
3. There are many apparent causes of the "problem" and they are inextricably tangled.
4. It is just not possible to be sure when you have the correct or best solution; there is no "stopping" rule.
5. The understanding of what the "problem" is changes when reviewed in the context of alternative proposed solutions.

These characteristics suggest why wicked problems are a major reason why the future becomes unknowable. There has been a stream (Camillus 2016) of literature on wicked problems that describes processes for dealing with extreme complexity and uncertainty. The recommendations on how to deal with unknowable futures when developing BoH Strategies are drawn from this stream of literature and the approaches we have observed in companies. They include the following:

1. Embracing "BoH" values
2. Envisioning a desired future
3. Adopting an actions-to-strategy sequence
4. Implementing real-time issue management
5. Engaging in rapid prototyping and experimentation

Embracing "BoH" Values

In order for BoH Strategies to succeed in an organization, its leadership must embrace a particular set of values. In describing these BoH values, a taxonomy of universal human values proposed by Shalom Schwartz (2012) is illuminating and helpful. This taxonomy, presented in Figure 3.3, helps illuminate values that are aligned with BoH Strategies. BoH values

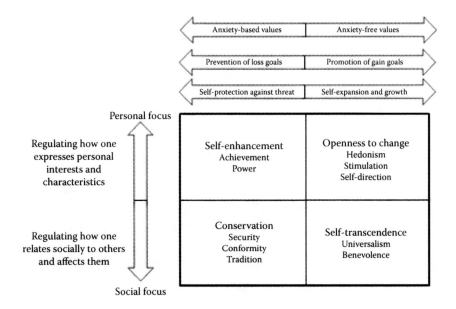

FIGURE 3.3

Dimensions of the universal values structure. (From Schwartz, S.H., 2012, An Overview of the Schwartz Theory of Basic Values, Online Readings in Psychology and Culture, 2(1). http://dx.doi.org/10.9707/2307-0919.1116.)

fall into the right-hand quadrants, which are described as being anxiety-free, supporting the promotion of gain rather than the minimization of loss, and oriented toward self-expansion and growth rather than self-protection against threats.

It is apodictic that organizational leadership that is anxiety free focused on promotion of gain rather than preventing loss, and seeking self-expansion and growth will be predisposed to adopt and implement BoH Strategies. Without such a predisposition, there is little incentive for managers subscribing to the conventional wisdom to consider anything beyond short-term, accounting profits that are commonly perceived to be the route to maximizing shareholder value. The discontinuities created by wicked problems that make the projection of accounting profits suspect at best are likely to be discounted by leadership that is insensitive to the complexities and uncertainties of today's business environment.

Schwartz's research bodes well for the future acceptance and implementation of BoH Strategies. In the vast majority of nations that he studied, benevolence, universalism, and self-direction were given the highest importance, while power, tradition, and stimulation received the lowest importance.

Envisioning a Desired Future

Classic planning processes and variations, such as MIT professor J.B. Quinn's (1980) logical incrementalism, employ visions and visioning as part of the process. These processes are based on the assumption that it is possible to predict future contexts. However, when faced with extreme complexity and uncertainty of the kind that spawns wicked problems, which effectively create an unknowable future, visions can no longer be informed and framed by an understanding of a likely business context. In such situations, where predictions and projections are futile, organizations may have no alternative but to envision a future that they would like—rather than expect—to have happen. There is truth in clichés, and the classic aphorism that has become a cliché in the global business environment—"when you can't predict the future, create it"—certainly applies here.

One can reasonably speculate on the likely constituent elements of an organizational vision in the context of an unknowable future. First, the desired future that an organization envisions will arguably be greatly influenced by its values. The organization's *core values* (e.g., sharing economic value added equitably among stakeholders, demonstrating simplicity in purpose and operations, caring for the environment) provide meaning and guidance to the vision.

Second, the organization's *enduring aspirations* (e.g., leadership in technology, innovativeness in value propositions, dominance in market share, growth significantly higher than selected market indices, empathetically serving customer needs, enhancing the quality of life at the bottom of the pyramid, synergistically integrating economic value and social benefit) are what will probably be employed to give substance to the vision.

Third, the organization's *distinctive competencies*, which essentially define the possibilities available to the organization, would both direct and constrain its vision. Of course, an ambitious management team that commits to BHAGs would very well seek to add to the array of competencies that the organization possesses. This could be accomplished by a variety of means such as training and development of existing personnel, recruiting personnel with the needed skills and knowledge, acquiring organizations that have demonstrated the needed competency, and forging alliances with organizations that possess complementary competencies.

Later, we will make the argument, which is amply supported in the literature (Camillus 2011), that values, aspirations, and competencies are the three components of the highly relevant and critically important construct of organizational "identity."

Adopting an Actions-to-Strategy Sequence

When faced with impenetrable ambiguity, resulting in part from the conflicting priorities of diverse stakeholders, another response relevant to the challenge was suggested by Charles Lindblom (1959). Lindblom identified the intriguing reality that while stakeholders may have different priorities or values, there are certain actions that all may agree are acceptable. For instance, governments, nongovernmental organizations, and firms, though their values, priorities, and motivations, may be different, yet may all agree that providing access to financial services to people at the bottom of the pyramid is a good thing.

Lindblom calls this a branch-to-root approach; it is the reverse of the classic approach, which is to proceed from strategy to actions, which Lindblom characterizes as root-to-branch. While Lindblom did claim that this approach is rooted in science, his labeling of the process as "muddling through," while arguably appropriate and evocative, did not contribute to the popularization of this powerful and insightful approach. Also contributing to the lack of widespread awareness and acceptance of this approach was the fact that Lindblom labored in the policy sciences vineyard, rather than in the business arena. However, an article that applied this approach to the strategic planning arena was authored by Robert H. Hayes (1985). He described this approach as "Strategic Planning—Forward in Reverse;" again a less than felicitous label.

When the future is unknowable, the notion of a set of acceptable actions leading to strategy is not entirely counterintuitive, though it is of course contrary to conventional wisdom. Another argument for building strategies from a foundation of actions is based on the concept of *robust actions*—actions that work effectively in multiple contexts. The hallmark of unknowable futures is that while no specific future may be identified in the context of the complexity and uncertainty caused by the incendiary mix of globalization and shared value, it may be possible to speculate on a variety of different scenarios that conceivably could emerge. Experience has shown that there is a common set of actions, *robust actions* (Fahey and Randall 1998), which make sense and contribute to added value in each and every one of the possible scenarios that are conjectured. Robust actions feed into Lindblom's branch-to-root, action-to-strategy sequence. We will present an approach to identifying robust actions in a later chapter.

Robust actions also serve as a foundation for a real-options approach that can strengthen a BoH Strategy plagued by chaotic ambiguity. Real options (Trigeorgis 1996) essentially require an organization to make the minimum investment necessary to enter and retain a foothold in a business opportunity or a strategic initiative, while gathering information that sheds more light on the context and the future. When greater clarity is achieved, a more confident determination can be made of whether and how to go ahead with the business opportunity or strategic initiative. Major investments and more defined commitments can then be made.

A real-options approach recognizes that the veils of uncertainty and the knots of complexity may lift and unravel over time to where a reasonable projection of a likely future may be made and resources may be committed in anticipation of this future happening. Robust actions enable organizations to move forward, while waiting for information that clarifies the future to the point where risks can be calculated and reasoned investment decisions can be made. The alternative is to be paralyzed with indecision or pressured into making major investments with inadequate information.

Implementing Real-Time Issue Management

Guided by its identity, motivated by its vision, and committed to robust actions, an organization that steps into an unknowable future may be as well positioned as possible, but unanticipated problems and sudden opportunities will present themselves. Annual planning processes conducted by most organizations cannot cope with the reality and inevitability of randomly emerging strategic issues. In place of episodic planning processes, a continuous issue-management (Camillus and Datta 1991) process is demanded by nature of an unpredictable future. Real-time issue management, as opposed to episodic strategic planning at preset intervals, is logically consistent with and supportive of "scientific muddling through," and the real-options approach.

Engaging in Rapid Prototyping and Experimentation

Concomitant with real-time issue management is the need to test the organization's responses to the emerging issues. It is important to learn as quickly as possible whether the responses are effective. A rapid prototyping approach to the new strategic initiatives and business models is of relevance here. The mindset is one of experimentation, of hypothesis testing, where the object is to learn and improve, as opposed to declaring failure or defeat.

In the context of unknowable futures, carefully planned initiatives are best viewed as *experiments*. Whether the initiative works well or not, it can provide valuable insights into cause–effect relationships and the validity of underlying assumptions. Even when unsuccessful, these experiments can serve to identify needed capabilities that the organization lacks, and inspire the development of those capabilities.

It is important to consider the relationship of existing capabilities and those identified as needed in order to effectively implement initiatives. The intent should be to develop an array of interrelated capabilities that reinforce one another. The conventional notion of a "core competency" that gives rise to multiple new products or services is not as meaningful or significant as a growing array of competencies that enable organizations to build on existing strategies and embrace BoH Strategies that meet different customer needs and venture into new and different markets.

Compiling the Responses to Unknowable Futures: Feed-Forward Processes

The approaches and techniques to respond to unknowable futures that were discussed are as follows:

1. Embracing BoH values and affirming the organization's identity—articulating what is core, enduring, and distinctive about the firm, which can transcend discontinuities and disruptions
2. Visioning—defining a future that would motivate and guide decision making
3. Actions-to-Strategy Sequence—the branch-to-root process or "scientific muddling through"—employing robust actions and a real-options approach to resource allocation
4. Real-Time Issue Management—continuous strategic planning triggered by issues as they emerge or appear on the horizon
5. Rapid Prototyping/Experimentation—testing the organization's responses to emerging issues, problems, and opportunities and building a growing array of capabilities

These responses to an unknowable future align with a feed-forward (Veliyath 1985) approach to planning and control. Feed-forward-based planning and control processes work from desired futures backward, as opposed to the classical feedback approach, which seeks to draw lessons from the past. We will discuss these processes in detail in Chapter 6.

REFERENCES

Brown, T. (2008). Design thinking. *Harvard Business Review*, 86(6): 84–92.

Camillus, J.C. (2008). Strategy as a wicked problem. *Harvard Business Review*, 86(5): 98–106.

Camillus, J.C. (2011). Organisational identity and the business environment: The strategic connection. *International Journal of Business Environment*, 4(4): 306–314.

Camillus, J.C. (2016). *Wicked Strategies: How to Conquer Complexity and Confound Competitors*. Toronto, Canada: University of Toronto Press.

Camillus, J.C., and Datta, D.K. (1991). Managing strategic issues in a turbulent environment. *Long Range Planning*, 24(2): 67–74.

Chen, H., Chiang, R.H.L., and Storey, V.C. (2012). Business intelligence and analytics: From big data to big impact. *MIS Quarterly*, 36(4): 1165–1188.

Christensen, C.M. (2013). *The Innovator's Dilemma: When New Technologies Cause Great Firms to Fail*. Boston: Harvard Business Review Press.

Collins, J.C., and Porras, J.L. (1996). Building your company's vision. *Harvard Business Review*, 74(5): 65–77.

Fahey, L., and Randall, R.M. (Eds.). (1998). *Learning From the Future: Competitive Foresight Scenarios*. New York: John Wiley & Sons.

Hayes, R.H. (1985). Strategic planning—Forward in reverse. *Harvard Business Review*, 63(6): 111–119.

Immelt, J.R., Govindarajan, V., and Trimble, C. (2009). How GE is disrupting itself. *Harvard Business Review*, 87(10): 56–65.

Kanter, R.M. (1994). Collaborative advantage: The art of alliances. *Harvard Business Review*, 72(4): 96–108.

Lindblom, C.E. (1959). The science of "muddling through." *Public Administration Review*, 19(2): 79–88.

Mehrpouya, H., Maxwell, D., and Zamora, D. (2013). Reflections on co-creation: An open source approach to co-creation. *Participations*, 10(2): 172–182.

Normann, R., and Ramirez, R. (1993). Designing interactive strategy. *Harvard Business Review*, 71(4): 65–77.

Pfitzer, M., Bockstette, V., and Stamp, M. (2013). Innovating for shared value. *Harvard Business Review*, 91(9): 100–107.

Porter, M.E. (2011). *Competitive Advantage of Nations: Creating and Sustaining Superior Performance*. New York: Simon and Schuster.

Quinn, J.B. (1980). *Strategies for Change: Logical Incrementalism*. Homewood, IL: Irwin.

Schwartz, S.H. (2012). An Overview of the Schwartz Theory of Basic Values. Online Readings in Psychology and Culture, 2(1). http://dx.doi.org/10.9707/2307-0919.1116

Strong, C. (2015). *Humanizing Big Data: Marketing at the Meeting of Data, Social Science and Consumer Insight*. London: Kogan Page.

Trigeorgis, L. (1996). *Real Options: Managerial Flexibility and Strategy in Resource Allocation*. Cambridge, MA: MIT Press.

Veliyath, R. (1985). Feedforward Orientation in the Strategic Management Process: A Contingent Choice (Doctoral dissertation, University of Pittsburgh).

4

The Business of Humanity Management Framework

INTEGRATING THE RESPONSES TO DISRUPTIVE TECHNOLOGIES, CONFLICTED STAKEHOLDERS, AND UNKNOWABLE FUTURES

The responses that we have compiled to the challenges of disruptive technologies, conflicted stakeholders, and unknowable futures constitute the elements of an integrated management framework. The sets of responses that were developed in response to these three challenges were compiled under the labels of innovative business models, co-creation of value, and feed-forward processes.

Innovative business models, co-creation of value, and feed-forward systems need to work in concert in order to maximize their effect. We will examine the elements of these three to detect common patterns, identify interactions, and link related actions and techniques in order to create an integrative framework that can effectively and efficiently support the development and deployment of Business of Humanity (BoH) Strategies.

The individual responses included under innovative business models, co-creation of value, and feed-forward processes are listed in Table 4.1. The commonalities and connections across these individual responses were identified, so related responses across the three challenges were clustered together. The clusters of responses that emerged constitute three distinct constructs that are the foundation of the BoH framework for strategic management. Table 4.1 presents the three constructs and the underlying responses.

TABLE 4.1

Constructs Underlying the BoH Management Framework

Macro-Response (Challenges) / Response Clusters	Innovative Business Models (Disruptive Technologies)	Co-Creation of Value (Conflicted Stakeholders)	Feed-Forward Processes (Unknowable Futures)
Construct 1 **Identity**	Incorporate social responsibility into the business model • Employ "humane" criteria for evaluation • Address "humankind" including emerging economies and low-income segments	Recognize and engage diverse stakeholders Manage stakeholders • Co-incentivize	Articulate identity—core values, enduring aspirations, and distinctive competencies
Construct 2 **Feed-forward**	Explore disruptive technologies and innovation along the entire value chain Employ design thinking and data analytics Build a dynamic array of capabilities	Engage in co-creation across the entire value chain Promote and manage alliances along strategic, project/venture, operational, interpersonal, and cultural dimensions	Engage in visioning Follow an actions-to-strategy sequence in planning • Robust actions • Real options Adopt real-time issue management • Continuous scanning and response
Construct 3 **Frugal engineering**	Relate empathetically to the customer Set price point–related BHAGs	Employ skunk works and superteams Nurture and connect innovation ecosystems • Cross-industry connections • Cross-market/country connections	Rapid prototyping and experimentation

Construct 1: Identity

The responses to the three challenges that are clustered to form this construct of "identity," listed in our understanding of a logical hierarchy, are as follows:

- Articulate identity—core values, enduring aspirations, and distinctive competencies
- Incorporate social responsibility into the business model
    - Employ "humane" criteria for evaluation
    - Address "humankind" including emerging economies and low-income segments
- Recognize and engage diverse stakeholders
    - Manage stakeholders
    - Co-incentivize stakeholders

Identity is the driver of the strategic posture and the touchstone for decision making in the organization. Each and every element of the construct illuminates the *raison d'être* of the firm and gives special meaning and substance to BoH Strategies. We will explore this construct in detail in Chapter 5.

Construct 2: Feed-Forward Processes

The responses incorporated in the construct of "feed-forward processes" are as follows:

- Engage in visioning
- Adopt real-time issue management
    - Continuous scanning and response
- Follow an actions-to-strategy sequence in planning
    - Robust actions
    - Real options
- Employ design thinking and data analytics
- Explore disruptive technologies and innovation along the entire value chain
- Engage in co-creation across the entire value chain
- Promote and manage alliances along strategic, project/venture, operational, interpersonal, and cultural dimensions
- Build a dynamic array of capabilities

Feed-forward processes are the organization's shield and spear in the face of complexity and uncertainty. They build on the foundations of the organization's competencies, guided by its values and directed by its aspirations.

Construct 3: Frugal Engineering

The "frugal engineering" construct includes the following responses:

- Relate empathetically to the customer
- Set price point–related BHAGs
- Employ skunk works and superteams
- Nurture and connect innovation ecosystems
  - Cross-industries connections
  - Cross-markets/countries connections
- Rapid prototyping and experimentation

Frugal engineering is a critically important construct. It supports key characteristics of BoH Strategies: global reach, focus on low-income segments, price point–driven BHAGs, and reverse innovation.

INTEGRATING THE CONSTRUCTS TO CREATE THE BoH MANAGEMENT FRAMEWORK

These three constructs comprise the set of instruments that managers can employ to create BoH Strategies. They are the components that constitute the management framework that supports BoH Strategies.

The three constructs are related and interactive. Identity is intimately connected to the other two constructs—feed-forward processes and frugal engineering. The connections are bidirectional. In the first sequence, the organization's identity outlines a desired future, shaped by its values and illuminated by its aspirations. Feed-forward explores and maps out the desired future—identifies the actions that create the future and the path that will lead to this future. Frugal engineering is directed toward markets and customers glimpsed through the lens of identity and mapped out through feed-forward processes. The capabilities and innovations, which derive from frugal engineering focused on serving these markets and customers, enrich the competencies that are part of the organization's identity.

The logic also holds in the reverse direction, where the aspirations and competencies that are part of the organization's identity enable and guide the developments in frugal engineering, which modify the screens employed in feed-forward processes, enabling strategic alternatives that can influence the aspirations espoused in the organization's identity.

The proposed BoH Management Framework is diagrammed in Figure 4.1.

The BoH Management Framework creates BoH Strategies. These strategies in turn create innovative business models, support co-creation of value, and forge a sustainable future. This was presented as the purpose of BoH Strategies in Figure 3.1 in Chapter 3 ("Responding to the Strategic Challenge"). For ready reference, this figure is reproduced as Figure 4.2, this time illustrating what BoH Strategies have now been shown to be capable of, namely, "Converting Challenges to Competitive Advantages."

Disruptive technologies that threaten the organization's viability form the basis for innovative business models, conflicted stakeholders are disarmed and engaged in co-creating value, and unknowable futures are transmuted into a desired sustainable future.

Chapters 5 through 7 will discuss identity, feed-forward processes, and frugal engineering with the details that are necessary to employ them effectively in developing and deploying BoH Strategies.

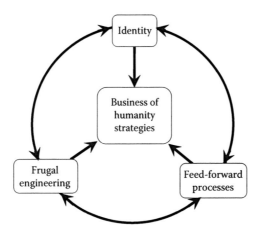

FIGURE 4.1
The BoH Management Framework.

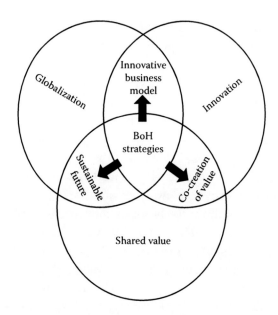

FIGURE 4.2

BoH Strategies: Converting challenges to competitive advantages.

5

Identity as Anchor, Beacon, and Compass

In order to explore the construct of "identity," let us recap the elements that were combined to constitute the construct:

- Articulating identity—core values, enduring aspirations, and distinctive competencies
- Incorporating social responsibility into the business model
  - Employing "humane" criteria for evaluation
  - Addressing "humankind," including emerging economies and low-income segments
- Recognizing and engaging diverse stakeholders
- Managing and co-incentivizing stakeholders

We will consider each of these elements individually, assessing their relationship with and influence on Business of Humanity (BoH) Strategies.

ARTICULATING IDENTITY

The concept of an organization's identity has been gaining much traction in recent years. Traditional definitions of the organization's *raison d'être*, such as mission, concept of business, and strategy no longer serve effectively as the touchstone for decision making in light of the sea changes in the business environment driven by technological developments, regulatory changes, demographic shifts, and changing social mores.

Discontinuities result from these drivers of change. Traditional competitive strategy, defined by the organization's product–market–technology (PMT) choices, does not fare well when faced by discontinuities. PMT statements exercise a strong and constraining influence on the important choices made by the organization. These choices tend to moor the firm to its existing business model, while the forces of globalization, innovation, and shared value demand major changes.

Mission statements and definitions of the organization's concept of business also tend to focus, like traditional strategy, on elements such as products, markets, and technology. Abell (1980), for instance, in one of the better-known works on defining the business, suggests that business definitions should include three dimensions (which just happen to map onto the PMT definition of strategy):

- Served Customer Groups (i.e., Markets)
- Served Customer Functions (i.e., Products/Services)
- Technologies Utilized (i.e., Technology)

In a world that is constantly changing because of myriad forces, these traditional ways of defining an organization are obsolete or have ephemeral meaning. Organizations need a construct that illuminates and guides decision making even through sea changes and discontinuities, a touchstone for decision making, which possesses the following characteristics:

- Clear definition of what is *core* to the organization and continues to be relevant and inviolable through radical changes in the environment
- Enduring substance and meaning, evocatively guiding decision making, transcending repositioning and transformation of the organization
- Insightful identification of what is *distinctive* about the organization, suggesting sources of competitive advantage

The three components of identity—(1) the *values* embraced by the organization, (2) the *aspirations* (Camillus 2008; Cyert and March 1963) that are espoused, and (3) the *competencies* (Prahalad and Hamel 1990) it possesses—display the sought-after characteristics. Values are at the *core* of the organization, determining which stakeholders are important, which strategies are attractive, what policies are important, and what actions are acceptable. The long-term goals and the vision of where the organization wishes to be are the *enduring* ideas captured by aspirations. They can transcend discontinuities and disruptions with continued meaning and relevance. The competencies possessed by the organization are its source of competitive advantage and awards make it *distinctive* and different from other organizations. Values, aspirations, and competencies transcend discontinuities and define the organization's identity.

Values: It has been asserted that firms are founded on a "logic of values" (Freeman et al. 2007) and that values are of "sublime importance" (Camillus 2011). Values, it can therefore be argued, inform and perhaps even determine the *raison d'être* of firms. Values can directly impact the strategy of the firm. Values determine the relative importance that the organization ascribes to its various stakeholders. Consider the three most commonly discussed stakeholders:* shareholders, customers, and employees. The importance given to these three classes of stakeholders is a strategic choice that the firm's leadership has to make (Simons 2010). This choice significantly impacts the organization's strategy and performance. The importance given to each of these stakeholders actually does depend substantially on the values espoused by the firm.

Of late, there has been growing recognition of the importance of employees in the quest to maximize profits. Both practitioners (Nayar 2010) and academics (Pfeffer 2010b) have described and explained how and why a focus on employees leads to both higher profits and greater sustainability. Milton Friedman (1970) argues alternatively that it is unethical for a firm not to give primacy, if not sole importance, to shareholders. Dean of Harvard College and Harvard Business School professor Rakesh Khurana (2007), on the other hand, makes a powerful argument that giving primacy, if not sole attention, to shareholder value delegitimizes management as a profession. Khurana argues that "It was only after a sustained quest for social and moral legitimacy—finally achieved through the linkage of management and managerial authority to existing institutions viewed as dedicated to the common good—that management successfully defined its image as a trustworthy steward of economic resources represented by the large, publicly held corporation" (Khurana 2007, p. 3). Values determine whose philosophy the firm adopts.

Who is right is not important to this discussion. What is important is to recognize that values impact strategy, and that this impact happens regardless of the industry in which the firm operates. Values are at the

---

* See, for instance, the foci of the "balanced scorecard" as discussed in Kaplan and Norton (1996).

core of the organization. Values transcend disruptions. Values can have a guiding influence on strategy and performance in other ways. Values can determine whether a firm employs BoH criteria such as environmental sustainability, gender equality, diversity, integrity, quality, and safety in choosing between strategic alternatives. Strategies that align with these humane criteria have an impact on the firm's performance. There is a growing body of research and a broadening acceptance by managers of the importance of these humane values and their positive impact on both the bottom line and economic sustainability (Camillus 2014). These values are core to the construction of BoH Strategies.

Regarding "human" values, perhaps the most critical notion in the BoH model is that layoffs should be eschewed whenever possible. In a cover story in *Newsweek*, Stanford University professor Jeffrey Pfeffer (2010a) presents the emotional and physical toll of layoffs and downsizing. Following a layoff, the employees that are retained experience a loss in morale and survivor's guilt. They also tend to seek alternative employment, resulting in the loss of some of the best and most capable survivors of layoffs. Pfeffer argues that layoffs not only result in morbidity and significant health issues among employees but also tend to have deleterious effects on the bottom line. The dysfunctional mix of survivor's guilt and damaged morale that accompanies layoffs has an inevitable, negative impact on profits.

Firms need to be sensitive to the implications of layoffs. Employees possess vital tacit knowledge that cannot be codified in manuals or included in databases. Consequently, the firm's source of sustainable competitive advantage probably resides with its employees. Too often, especially in U.S. firms, management teams respond to pressures for cost reduction by laying off employees. While labor laws permit such layoffs in the United States, firms need to keep in mind that the capabilities that are the source of competitive advantage often reside with the employees.

Aspirations: Cyert and March's (1963) seminal work on the behavioral theory of the firm provides a great perspective on aspirations. They demonstrated that a higher degree of stretch or reach in an organization's aspirations can motivate a more concerted search for innovative alternatives, hence affecting the organization's performance. This affirms the earlier assertions about the positive impact of stretch goals and BHAGs on innovation.

Recognizing the firm's responsibility toward multiple stakeholders articulated by Freeman et al. (2007), its aspirations should encompass goals beyond profits. And as we noted earlier, Khurana (2007) makes a powerful argument that giving sole attention to shareholder value delegitimizes management as a profession. Also, the response to shared value requires a commitment to multiple goals. At the simplest level, the "balanced scorecard" (Kaplan and Norton 1996) which has achieved wide recognition and acceptance, suggests that in addition to shareholder-oriented financial goals, the firm needs to address goals that are responsive to the customer, that relate to processes and productivity, and that reflect the firm's growth and innovation. This set of goals will certainly color the organization's strategy. In addition to recognizing stakeholders other than shareholders, the balanced scorecard approach does make an important point: the firm needs to seek to improve the performance of the existing business and simultaneously seek to exploit innovations that could lead to new businesses. Wicked Strategies need to meet both of these goals—improving the existing business and growing into new businesses.

The transformational growth that Wicked Strategies seek also requires goals that are not constrained by the past. For instance, market share, which is perhaps the most widely employed measure of strategic performance and competitive strength, may not be suitable for incorporation in the organizational Identity. Market share is substantially a historically oriented measure in that it refers to a historical conception of the market. It focuses management attention on traditional definitions of the market rather than motivating managers to anticipate, embrace, and stimulate disruptive change. An alternative, enduring goal derived from a desired future would be, for instance, the proportion of revenues and profits that come from new markets each reporting period.

In short, aspirations need to incorporate the following:

- Multiple goals responding to the priorities of significant stakeholders
- Future-derived goals that motivate the firm to embrace change
- Stretch goals that push performance beyond the existing business

Competencies: This is the third constituent element of the identity construct. The importance and lasting nature of competencies have been brought home by the growing emphasis on the resource-based view of the organization (Barney 1991). There have been several dissertations honing the concept of competencies. For instance, the notion of sustainable competitive advantages (Ghemawat 1986) such as the ability to learn faster than the competition has been proposed. The business process-oriented variation of competency that has been labeled "capabilities" (Stalk et al. 1992) has also gained much attention. But, the most compelling and widely accepted concept related to competency is that of core competency proposed by Prahalad and Hamel (1990). While there is great value and appeal to the idea of core competency, the approach to competencies that is proposed in the context of identity is significantly different from the popular understanding of core competency. Instead of a focus on a core competency, we propose that an array of interrelated competencies is necessary to identify, create, and implement BoH Strategies. We will elaborate on the proposed array of competencies in the following chapter on the feed-forward construct.

INCORPORATING SOCIAL RESPONSIBILITY

In Chapter 3, we discussed that BoH Strategies demand and are driven by a particular set of values. For BoH Strategies to be formulated and implemented by organizations, it is critically important that the values of the organization embrace social responsibility.

The BoH Proposition espouses the two dimensions of humanity—humaneness and humankind. BoH Strategies, which are built on *humane* considerations, should logically value environmental sustainability, gender equality, diversity, integrity, quality, and safety in choosing between strategic alternatives. Strategies that align with these humane criteria have an impact on the firm's performance. There is a growing body of research and broadening acceptance—as discussed in Chapter 1—by managers of the importance of these humane values and their positive impact on both the bottom line and economic sustainability (Camillus 2014). BoH Strategies can also be expected to value *humankind*, which means valuing a global perspective and all segments of the market. Humaneness and humankind are important values that BoH Strategies must embrace in order to promote synergy between social and economic benefits.

RECOGNIZING AND ENGAGING DIVERSE STAKEHOLDERS

The previous discourse on the importance of humaneness and humankind as values makes the case for recognizing and engaging diverse stakeholders. Clearly, a singular focus on shareholders is contrary to the spirit and character of BoH Strategies. BoH Strategies, with their focus on social responsibility, have to go beyond the typical stakeholders, which include employees, existing customers, and shareholders. In terms of customers, there is the expectation that BoH Strategies will seek out new customers whose needs are not being met. Societal needs such as environmental sustainability and supporting the social compact, which are very much of concern to BoH Strategies, suggest that society and government, as well as NGOs that share such societal concerns, are stakeholders to be recognized.

MANAGING AND CO-INCENTIVIZING STAKEHOLDERS

This final element of the identity construct is a natural corollary of the recognition and engagement of diverse stakeholders. In order to engage diverse stakeholders, who inevitably will have a variety of different priorities and concerns, it would be necessary to adopt aspirations that, if not aligned with these priorities, are at least not in conflict with them. In addition, the activities that support BoH Strategies have to be acceptable to the stakeholders, and economic value added has to be equitably shared.

Identity serves as the foundation and a powerful stimulus for BoH Strategies. The values that are part of the identity construct serve as the anchor, if you will, that enable organizations to withstand the maelstrom created by extreme uncertainty and complexity. Aspirations serve as the beacon that beckons the organization toward a desired and sustainable future. And competencies can function as a compass that guides the organization through a maze of strategic alternatives.

How the identity construct affects and can be purposefully employed to generate and evaluate strategic alternatives that form BoH Strategies will be discussed in Chapter 8.

REFERENCES

Abell, D.F. (1980). *Defining the Business: The Starting Point of Strategic Planning*. Englewood Cliffs, NJ: Prentice-Hall.

Barney, J. (1991). Firm resources and sustained competitive advantage. *Journal of Management*, 17(1): 99–120.

Camillus, J.C. (2008). Strategy as a wicked problem. *Harvard Business Review*, 86(5): 98–106.

Camillus, J.C. (2011). Organisational identity and the business environment: The strategic connection. *International Journal of Business Environment*, 4(4): 308.

Camillus, J.C. (2014). The business case for humanity in strategic decision making. *Vilakshan*, 11(2): 141–158.

Cyert, R.M., and March, J.G. (1963). *A Behavioral Theory of the Firm*. Englewood Cliffs, NJ: Prentice-Hall.

Freeman, R.E., Harrison, J.S., and Wicks, A.C. (2007). *Managing for Stakeholders: Survival, Reputation and Success*. New Haven, CT: Yale University Press. p. 6.

Friedman, M. (1970, September 13). The social responsibility of business is to increase its profits. *The New York Times Magazine*, The New York Times Company.

Ghemawat, P. (1986). Sustainable advantage. *Harvard Business Review*, 64(5): 53–58.

Kaplan, R.S., and Norton, D.P. (1996). Using the balanced scorecard as a strategic management system. *Harvard Business Review*, 74(1): 75–85.

Khurana, R. (2007). *From Higher Aims to Hired Hands: The Social Transformation of American Business Schools and The Unfulfilled Promise of Management as a Profession*, Princeton, NJ: Princeton University Press.

Nayar, V. (2010). *Employee's First, Customers Second*, Boston, MA: Harvard Business Press.

Pfeffer, J. (2010a). Lay Off the Layoffs. *Newsweek*, February 15, 32–37.

Pfeffer, J. (2010b). Building sustainable organisations: The human factor. *Academy of Management Perspectives*, 24(1): 34–46.

Prahalad, C.K., and Hamel, G. (1990). The core competency of the corporation. *Harvard Business Review*, 68(3): 79–91.

Simons, R. (2010). Stress-test your strategy: The seven questions to ask. *Harvard Business Review*, 88(11): 93–100.

Stalk, G., Evans, P., and Shulman, L.E. (1992). Competing on capabilities: The new rules of corporate strategy, *Harvard Business Review*, 70(2): 57–69.

6

Feed-Forward to a Visionary Future

Traditional strategy development relies substantially on feedback—which means learning from experience and analyzing actual performance in relation to planned performance. Disruptive technologies and unknowable futures delink the future from the past, making traditional strategy development processes inadequate. Feed-forward (Veliyath 1985) processes and techniques address the disconnect that exists from the past by guiding management (Camillus 2015) in making choices today by working back from an anticipated or desired future, without necessarily relying on past experience.

In order to explore the construct of "feed-forward processes" proposed as a response to disruptive technologies, conflicted stakeholders, and unknowable futures, let us first recap the elements that constitute the construct:

- Engaging in visioning
- Adopting real-time issue management
- Following an actions-to-strategy sequence in planning
- Employing design thinking and data analytics
- Exploring disruptive technologies and innovation along the entire value chain
- Engaging in co-creation across the entire value chain
- Promoting and managing alliances along strategic, project/venture, operational, interpersonal, and cultural dimensions
- Building a dynamic array of capabilities

Each of these elements in the construct are discussed below.

VISIONING

Visioning is the process by which the firm determines what and where it wishes to be in the future and also helps identify how to reach this goal. Visioning essentially fleshes out and operationalizes the organization's aspirations, which are part of the identity construct. Logically, there are three steps involved in a basic visioning process:

1. Visualizing and describing the *desired future* of the organization and its environment
2. Identifying the *"enablers"* that can bring about or result in the desired future
3. Developing an *implementation plan* for influencing or acquiring the enablers that can create the desired future

The example of a graduate school of business provides an illustration of the process. Graduate business education is big business and traditionally has served as a reliable source of cash for universities. But in recent years, graduate business education has descended into turmoil. There has been a worldwide proliferation of business schools. New players, including the governments of emerging economies and for-profit businesses, have been aggressively entering the arena. Businesses that need management talent are increasingly developing the talent in-house. Some others are shifting from recruiting graduates of master's programs to recruiting individuals with bachelor's degrees. The role and profession of management is being intensely debated. Distinguished academics such as Harvard professor Rakesh Khurana argue that the profession has lost relevance and credibility. Leading CEOs, such as Paul Polman of Unilever, are expressing their support for business models and management philosophies that align with Business of Humanity (BoH) values. Others are doubling down on the popular understanding of Milton Friedman's singular focus on profits and the shareholder.

Importantly, disruptive educational and distance-learning technologies are blurring the future and the viability of the vast majority of business schools. Harvard Business School gurus Michael Porter and Clayton Christensen have engaged in public, intense debate about how their school should employ these technologies. The technology challenge is made more complex by the widespread emphasis on live human interactions and experience-based learning as the key to developing leadership skills and effective managers.

And in the midst of this philosophic, pedagogic, technological, and, indeed, existential turmoil, applications to MBA programs have been declining significantly. In this turbulent context, a hypothetical North American graduate management school may execute the following strategy based on our three-step process: vision, enablers, and implementation. The *vision* expresses the enduring values, aspirations, and competencies, the elements of the identity construct, of the organization. The *enablers* derive from the vision. Each element of the *implementation plan* is supported by and also reinforces one or more of the enablers. Both the organization and its context are targeted for transformation by the vision and the implementation plan. In the context of the graduate school, the synergy of the three elements would look like this:

1. **Vision:** Become an acknowledged world-class, innovative, management school:
   a. Renowned for a revolutionary, pragmatic, experience-based, career-oriented pedagogical approach that adds high and recognized value to the students' capabilities.
   b. Respected for contributing to management theory and practice with a global perspective and reach.
   c. Prized as an ally that enhances both the economic value and social benefit generated by its partners.
2. **Enablers:**
   a. Global, sustained media recognition for radical and effective approaches to preparing students, which are advocated and validated by leading executives and academics.
   b. Intense and intimate partnerships with selected, leading, corporate, nonprofit, governmental agency, media, and nongovernmental organization (NGO) employers of MBA graduates in multiple countries.
3. **Implementation Plan:**
   a. Share the vision, implementation plan, and economic and social sustainability projections with key stakeholders and donors, and build a war chest.
   b. Use the war chest to recruit two or three faculty renowned for both research and consulting, or institutional leadership—public scholars and visionaries of the stature of Paul Krugman, Larry Summers,

Michael Porter, Jeffrey Pfeffer, Jeffrey Immelt, Paul Polman, Elon Musk, Gary Hamel, Indra Nooyi, or Clayton Christensen.

c. Identify and partner with two leading management schools, one in Europe and one in Asia. (In Europe, preferably the United Kingdom, Switzerland, Holland, Denmark, or Norway; and in Asia preferably in India, so as to reduce language problems.)

d. Build intense and intimate relationships with select and committed corporate, nonprofit, governmental agency, media, and NGO partners along the dimensions of
   i. Curriculum and pedagogy development
   ii. Recruitment and selection of students
   iii. Scholarships
   iv. Internships
   v. Placement
   vi. Continuing education
   vii. Research projects

e. Design and implement a curriculum that is suited to students with three-plus years of managerial experience, with the following design parameters:
   i. Concentrated, three-module, experience-based program, completed in nine months to a year
   ii. First two modules to be taken in continents other than the student's home, with the third module in the home continent
   iii. Second and third modules to follow a cooperative education model with internships that are technically oriented in the second module and strategically oriented in the third module

f. Make extensive use of synchronous online education with
   i. An amphitheater-styled classroom in each of the three schools
   ii. Two large screens (with Cisco Telepresence or equivalent) in each classroom
   iii. Virtual teams for assignments, drawn from the three locations
   iv. Best faculty from the three schools

g. Include selective consulting for corporate, nonprofit, governmental agency, media, and NGO partners in the job description of faculty.

There are certain guidelines that can improve the effectiveness of the process. First, the enablers, of course, involve the transformation of

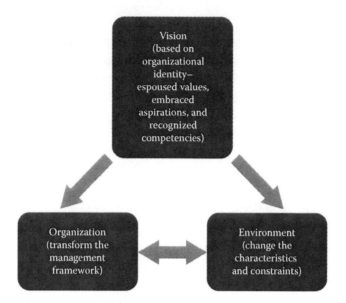

FIGURE 6.1
The visioning model.

the firm to achieve the desired mission. But it must be emphasized that enablers may also involve transforming the business environment. In the above example, for instance, educators and relevant government agencies need to be informed and convinced that experience-based learning and globalism are essential for effective management education.

The second guideline is that visioning must align with the firm's identity. The identity construct was created to provide strategic guidance for the firm, guidance that would transcend disruptions. Identity provides the platform on which the vision is built. The organization's mission must align with the BoH values incorporated in the identity construct. The recommended approach is diagrammed in Figure 6.1.

Changes in the management system that align with feed-forward and that support visioning are discussed below.

REAL-TIME ISSUE MANAGEMENT

Traditionally, strategy development processes occur on an episodic basis. Strategic planning usually takes place annually or triennially. However, the issues that demand a strategic response from the organization do not occur at these regular intervals. Assessing the environment at periodic intervals of a year or longer to determine emerging issues that need to be responded to would be ineffective, if not dangerous. Scanning the environment needs to be a continuous process. Strategic issues need to be addressed as and when they occur; not months or years after they emerge. Planning in response to emerging issues needs to be triggered as soon as significant developments are recognized; not wait for the next scheduled annual or triennial time.

Environmental scanning, in the context of developing BoH Strategies, is critically important. Potential sources and indicators of disruptive change in technology, regulations, and social mores and expectations need to be monitored. Unserved markets, where potential customers' basic needs are not being met, need to be identified.

Real-time issue management is essential in the context of innovative business models and related BoH Strategies. Issue management should be guided by the aspirations and visions that are part of the constructs of identity and feed-forward processes. The organization's identity provides the lenses and the filters for identifying which issues are of strategic importance, and which alternatives are best suited to responding to them. For instance, the process of "logical incrementalism" (Quinn 1980), mentioned in Chapter 3, is a tested and proven approach that envisages responses to issues as they emerge, with consistency in the responses being insured by commitment to a vision of where and what the organization desires to be.

ACTIONS-TO-STRATEGY SEQUENCE IN PLANNING

We discussed the relevance, importance, and utility of adopting an actions-to-strategy sequence in planning in Chapter 3. The two supporting concepts of "robust actions" and "real options" need to be borne in mind. The development of robust actions and the implementation of a real options approach are presented below.

The concept of *robust actions* derives from the realization that uncertainty can take the form of the existence of a number of possible futures (Courtney et al. 1997). This kind of uncertainty does not lend itself to typical forecasting techniques, which focus on identifying the future value of a variable in an anticipated future. Classic scenario planning also focuses on describing a likely future, with optimistic and pessimistic variations of this future. Multiple alternative futures require that a variety of possible scenarios be developed and that BoH Strategies responsive to each of the scenarios be formulated. Strategic actions that are common to all the strategies for responding to the various scenarios are, understandably, called "robust actions." The process (Dong-Gil Ko and Camillus 2001) of developing "possibility scenarios" and identifying robust actions is described below.

The process of developing possibility scenarios starts with the identification of key uncertainties that can affect the future. These uncertainties are employed to describe alternative futures/scenarios that may emerge. For each of the scenarios, strategies and action plans are developed. The strategies and related actions that are common to all the scenarios can then be identified. These strategies and actions that are common to all the scenarios constitute the robust set that can immediately be implemented. The environment is then continuously scanned for indications or signals about which of the possible scenarios that were identified is likely to happen. When it is clear that a particular scenario is emerging, the firm can then confidently invest in the comprehensive set of strategies and actions that were developed earlier for this scenario.

An example should help clarify the process. The current (2016) healthcare situation in the United States with the ongoing lawsuit regarding subsidies offered through the federal rather than state exchanges; the executive action taken by President Obama relating to the Affordable Care Act ("ObamaCare"); the pragmatic modifications made to ObamaCare during the course of implementation; the modifications imposed by the Supreme Court rulings, with the possibility of more modifications to come; the NGOs' and state legislatures' attempts to influence the healthcare choices for women; the breakthroughs in medical technology and new drugs; and the changing attitudes of the U.S. population, all combine to create an abundance of possible but unpredictable futures for the healthcare industry. In the case of ObamaCare, there are two key uncertainties that capture or reflect the aforementioned factors mentioned:

- The level of competition among providers of healthcare, as the size and economics of the healthcare market change, that is, "Provider Competition"
- The extent to which consumers, rather than employers or government agencies, will be responsible for paying for the services, that is, "Payment Shift to Consumer"

In juxtaposing the two possible states of each of these two uncertainties, as shown in Figure 6.2, four *possible* scenarios emerge. These four scenarios are labeled as follows:

1. Distress (significant increase in Provider Competition and significant increase in Payment Shift to Consumer)
2. Arms Race (significant increase in Provider Competition and status quo in Payment Shift to Consumer)
3. Getting-By (status quo in Provider Competition and status quo in Payment Shift to Consumer)
4. Belt Tightening (status quo in Provider Competition and significant increase in Payment Shift to Consumer)

For each of these four scenarios, detailed strategies and action plans need to be developed. The sets of strategies then need to be examined to identify for actions that are common to all four scenarios. These robust actions for U.S. health systems are as follows:

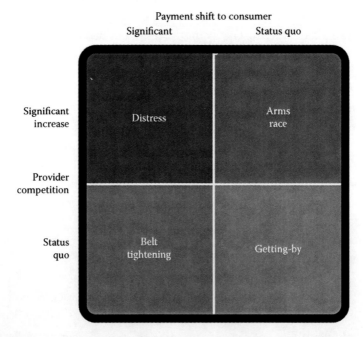

FIGURE 6.2
Possibility scenarios: U.S. health systems.

- **Building the brand image of the health system.** This would reduce the inclination of consumers to shift to competitive providers.
- **Providing wide access to measures of the quality of services.** This would enhance the brand equity and promote the confidence of consumers in the system.
- **Emphasizing customer service.** Consumer loyalty would naturally increase as a consequence of excellent service.
- **Investing in information technology.** Information technology and systems are key to supporting operations, enhancing and tracking quality, and supporting customer service.
- **Engaging in research.** This would improve the quality of patient outcomes and provide a competitive edge.
- **Constraining costs.** The ability to charge less would be critical in all contexts, and particularly so when patients exercise care in their choices because of a greater financial impact on them.

We discussed the concept of *real options* in Chapter 3. Real options (Trigeorgis 1996) essentially require an organization to make the minimum investment necessary to enter and retain a foothold in a business opportunity or a strategic initiative, while gathering information that sheds more light on the context and the future. When greater clarity is achieved, a more confident determination can be made of whether and how to go ahead with the business opportunity or strategic initiative. Major investments and more defined commitments can then be made. Robust actions also serve as a foundation for a real-options approach as they represent the initial investment to be made while awaiting clarification on which of the possibility scenarios is going to happen.

DESIGN THINKING AND DATA ANALYTICS

Design thinking starts with an empathetic understanding of the customer's needs. Professor Jhunjhunwala's experience, discussed earlier, reinforces the critical importance of such an understanding. In his case, it led him to shift from a competence-based focus on the technology of communication to its use in meeting the health, education, and livelihood needs of India's low-income population. In fact, in all the examples presented in Chapter 3, including TeNet, Arvind, GE, Tata, Vodafone CZ, and CavinKare, insightfully identifying with and understanding the customer was a common and critically important element in the process of motivating disruptive technologies and building innovative business models.

In addition to empathizing with the customer, design thinking, as described by Jon Kolko (2015), involves developing models that simplify and promote comprehension of the problem, testing and refining possible solutions, accepting uncertainty, and tolerating risk. The experimental, prototyping-based process espoused by design thinking is analogous to the approach to product innovation that employs an experiential process involving improvisation, real-time adaptation, and flexibility. This experimental approach to product innovation was found by Eisenhardt and Tabrizi (1995) to be more effective than rational, compression models in accelerating adaptive processes in organizations. The design thinking process as described by Kolko (2015) supports the creative generation of BoH Strategies.

Data analytics is the offspring of the information revolution—digital technology, improved statistical analysis, and massive computing power. Often used in the marketing arena, it aids in the effort to develop "customer insights." Such analysis cannot readily lead to the development of BoH Strategies because the products and services offered under these strategies tend to be substantially different from those existing, and the customers served tend to belong to very different market segments. Nevertheless, relationships that are revealed by analyzing available data may help develop hypotheses about cause–effect relationships that can be tested. More importantly, once a BoH Strategy has been implemented, and sufficient data have been gathered, data analytics can be of significant value. We see the design approach and data analytics as complementary, with the design approach helping to develop BoH Strategies and data analytics helping them evolve more efficiently.

Embracing design thinking requires both creativity and courage. But the potential rewards are substantial. The empathetic aspects of design thinking and the insights developed by data analytics can illuminate BoH Strategies pre- and post-implementation.

INNOVATION ALONG THE ENTIRE VALUE CHAIN

Within industries that have been disrupted, the organizations that have survived and start-ups that have been successful by and large demonstrate a common pattern in that disruptions, strategic responses, and innovations are not restricted to just technology, or product design and characteristics. The successful transformational strategic moves they employ affect or consider multiple elements in the value chain.

For instance, Vodafone CZ's explicitly articulated BHAG of "transforming the telecommunications industry" would have seemed like a pipe dream at best, if it had sought to transform the telecommunications industry in terms of technology and products. Vodafone CZ was a relatively small telecommunications company in the Czech Republic, with no significant R&D or technology development capabilities. The way Vodafone CZ planned to transform the industry was through the quality and kind of services it offered to individual customers. The overriding strategic direction adopted by Vodafone CZ was to "break all the rules for the customer." Vodafone CZ required no contracts, applied no switching penalties, empowered customer representatives to offer any amount of monetary compensation for shortcomings to customers, and offered existing customers all the incentives being offered to gain new customers. By focusing on the service element of the value chain, this small company essentially transformed the phone and Internet services to individual customers in the Czech Republic.

Similarly, CavinKare, a small start-up in India, beginning with an investment of $300, became a multi–hundred million dollar company in a few years by innovating multiple elements in the value chain of the toiletry industry. For instance, it packaged shampoo in single-use sachets that were affordable to customers at the bottom of the pyramid, added long-lasting, intense floral fragrances that suited the taste and bathing practices of its customers; distributed its product through novel outlets such as bicycle rental shops; offered complementary products with similar characteristics; and introduced promotions that rewarded customers for using shampoos—even those of competitors!

The disruptions that have been experienced by the imaging, publication, and music industries have taken place along multiple elements of the value chain. The traditional value chain in the imaging industry, for instance, is diagrammed in Figure 6.3.

FIGURE 6.3
Traditional (silver-halide) imaging industry value chain.

In traditional, silver-halide-based photography processes, mechanical devices (i.e., cameras) focused light reflected off the image being photographed for the appropriate period to trigger a chemical reaction. This constituted the *acquisition* step in the value chain. In the *processing* step, chemical reaction was stabilized. The stabilized image was then transferred to a transparent film and printed on paper for storage. In the *reproduction* step, when copies of the image were required, the images on the transparent film (i.e., negative) were reproduced on paper. Every element in this industry value chain was fundamentally transformed or modified by the advent of digital technologies for imaging and the possibilities created by the Internet. These transformational changes resulted in the value chain diagrammed in Figure 6.4.

The traditional imaging value chain has been fundamentally altered by the emergence of digital imaging replacing the silver-halide, chemical process. Coupled with the advent of the Internet, additional elements and changes characterize the now transformed imaging-industry value chain. The acquisition of the image is done by a charge-coupled device that converts the light, affecting it into digital signals that are processed for storage or display. The absence of a film on which the light images are gathered

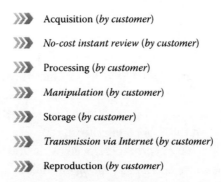

FIGURE 6.4
Transformed (digital) imaging industry value chain.

and then developed means that the image can be instantly checked by the person taking the photo. This does not cost anything because there are no tangible, physical media being employed. Furthermore, with appropriate software such as Adobe Photoshop, the image can be modified in any way that the image taker wants. Storage in digital form can be on a variety of storage media including the Cloud. The image can be made instantly available to anyone anywhere in the world through the Internet. It can be displayed or reproduced in a wide variety of ways.

There are two fundamental characteristics of the new value chain that have been altered. First, the cost structure has changed. Once the initial capital investment has been made, no additional costs for media or for displaying the image need to be incurred. Second, the customer is in complete charge of the entire value chain.

These two changes provide the basis for developing new business models that are aligned with BoH Strategies. The low-cost characteristic can be a basis for providing imaging capability and products to low-income consumers, even those at the base of the pyramid. Second, the ability for the customer to control all the elements of the value chain enables a business model that can empathetically respond to customer preferences.

In addition to these two changes, the capabilities offered by the Internet can provide an even more effective set of services that can delight the customer. What can provide more delight than for grandparents located 10,000 miles away to see an image of a new grandchild a minute or two after she or he is born. While not at the same level of significance, the ability of an image taker to modify his or her image, using photo editing capabilities, is a powerful response to a human need.

There are two generalizations that can be drawn from the example of the transformed imaging value chain. First, every element of an existing value chain offers the possibility for transformation that can support innovation in the business model and enable BoH Strategies. Second, additional elements may emerge in a transformed value chain, or existing elements can be eliminated. These changes again offer possibilities for new business models and BoH Strategies.

A third generalization, which digital imaging supports, but which is not nearly as apodictic as the previous two generalizations, is that providing a "systems solution" to a customer can more effectively serve the convenience and control needs of customers.

CO-CREATION ACROSS THE ENTIRE VALUE CHAIN

This digital-imaging industry value chain discussion highlights the potential for complementary products and technologies such as for processing, storing, and reproducing images. It is quite likely that such complementary products and technologies may not be possessed by firms that were operating on the basis of the traditional value chain. This reality surfaces another important consideration. Innovative business models that support BoH Strategies, not only may be based on changes in any element of the value chain, but might require new competencies that need to be developed either in-house or through alliances with organizations possessing the needed new competencies. Value can only be created in this new context by co-creating the model with partners who can provide the needed new competencies. Value added will have to be shared equitably with the providers of these complementary products and services, and the new competencies.

A corollary of this is that the traditional default condition of being in competition with other firms no longer necessarily holds. The emergence of complementary products and services, the need for developing or acquiring new competencies, and the imperative of empathetically serving the customer all suggest a mindset of "co-opetition" (Brandenburger and Nalebuff 1996). The hitherto dominant five-forces model of competition proposed by Michael Porter (1980) now takes a subsidiary position relative to the value net proposed by Brandenburger and Nalebuff (1996). The BoH imperative of "shared value" holds sway here.

Finally, a set of value-adding actions that respond to the changes in the value chain can be, if they are aligned and mutually reinforcing, melded into a strategy that exploits the disruption. Such an action-to-strategy sequence, Lindblom's (1959) branch-to-root process, is an effective approach to dealing with transformed value chains.

ALLIANCES

The alliances and partnerships that are necessary to cope effectively with transformed value chains need to be sensitively and purposefully managed to support BoH Strategies. They have been discussed in Chapter 3, so they are not elaborated upon here, except to point out that they are an important binding element that supports the preceding elements of the construct of feed-forward. Alliances essentially are mechanisms for co-creation and an enabler of innovative business models. Alliances are also an important means of accessing needed competencies.

DYNAMIC ARRAY OF CAPABILITIES

Competencies and the necessity for an array of competencies have been recognized and discussed in Chapters 3 ("Responding to the Strategic Challenge"), 4 ("The Business of Humanity Management Framework"), and 5 ("Identity as Anchor, Beacon, and Compass"), and of course earlier in this chapter ("Feed-forward to a Visionary Future").

The fundamental importance of competencies in the context of strategy has always been recognized. Perhaps the most widely accepted notion is that of the "core competency" (Prahalad and Hamel 1990). In relation to BoH Strategies, however, the idea for core competency that is the basis for new products and services and evolving strategies is less important and subservient to the idea that the organization needs a dynamic, constantly growing array of competencies. Innovation and innovative business models are an essential basis for BoH Strategies. Innovation can result in new competencies, and innovative business models often require new competencies.

This array of competencies has to be mutually supportive. Organizational sustainability demands that the competitive advantages deriving from existing competencies should not be discarded. New competencies that are related to existing competencies are more desirable than entirely new and unconnected capabilities. Furthermore, it would obviously be desirable if the new and existing competencies support and strengthen one another.

Marks & Spencer is a British department store that thrived for decades because of its commitment to quality and value of its products and its focus on caring for its employees. Its travails in recent years can be attributed to the fact that it did not develop or acquire new competencies to address the challenges and opportunities offered by globalization and new technologies. On the other hand, Walmart's incredible success can be attributed in part to its always growing new competencies that related to its foundational logistics, sourcing, and merchandising capabilities. In recognizing and responding to the imperative of globalization, Walmart also developed competencies in acquiring and merging retail organizations operating in very different contexts, responding to different cultural contexts, developing new business models relevant to the characteristics of each of the countries in which it operated, and in government relations.

BoH Strategies require firms to go beyond adjacent products or markets, to address low-income segments and markets in different countries. In support of this approach, Chakravarthy and Lorange (2008) convincingly argue that both growing existing businesses and entering new markets are necessary for profitable growth. Clearly, an array of competencies offer more points of leverage for entering new markets. Additionally, they

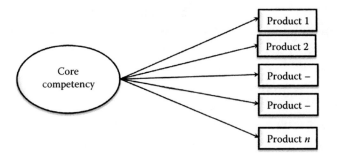

FIGURE 6.5
Conventional approach to competencies.

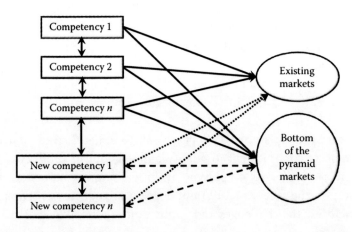

FIGURE 6.6
BoH Strategies' approach to competencies.

make the vital point that entering new markets will require the development of new competencies. Also, building on the argument of Hart and Christensen (2002) that meeting the needs of the base of the pyramid results in the development of disruptive technologies, one can expect that BoH Strategies will also help grow new competencies.

The conventional wisdom regarding competencies is presented in Figure 6.5. The approach to competencies in the context of BoH Strategies is presented in Figure 6.6.

THE FEED-FORWARD MINDSET

The feed-forward construct that helps build, implement, and evolve BoH Strategies creates a managerial mindset that is quite different from the conventional. Traditional management decision processes rely substantially on feedback—which means learning from experience and analyzing actual performance in relation to planned performance. BoH Strategies are quite different from those that are conventionally employed. The innovative business models and the constant co-creation of value along the entire value chain makes relying on an understanding of the past a quite inadequate basis for effective decision making. The new models that result from co-creation of value could make the disconnection even greater. Feed-forward (Veliyath 1985) processes and techniques address the disconnect that exists from the past by guiding management (Camillus 2015) in making choices today by working back from an anticipated or desired future, without necessarily relying on past experience.

While similar planning and control techniques may be employed in feedback as well as feed-forward approaches, there are profound differences between the two. Feedback is basically an exercise in remediation—correcting, learning, and improving performance in an existing business. On the other hand, feed-forward focuses on fashioning a future—a future that may be unrelated to the past—that the firm wishes to see happen. Feedback essentially involves performance appraisal and learning, while feed-forward focuses on managing uncertainty and an unknowable future. Feedback continuously improves managerial decision making by advancing the understanding of cause–effect relationships, while feed-forward involves a leap of faith, committing to a belief in a cause–effect

TABLE 6.1

Feed-Forward and Feedback

System Characteristic	Feed-Forward	Feedback
Purpose	**Fashioning the future**	**Remediation**
Focus	Managing uncertainty and an unknowable future	Performance appraisal and learning
Ethos	Committing to an assumed cause–effect relationship	Improving known cause–effect relationships
Trigger	Emerging issues (continuous)	Time period (episodic)
Information base	Predominantly future (possible scenarios)	Primarily past (historical data)
Analytical orientation	Visioning	Past performance

relationship in order to make strategic decisions and allocate resources. Feedback-oriented systems engage in episodic reviews of performance at specified intervals of time—say, monthly, quarterly, or yearly, while feed-forward systems trigger analysis whenever assumptions that have been made appear to be mistaken and new issues are spotted. Feedback employs databases that collect historical data, while feed-forward works with information and insights derived from possible future scenarios. In short, feedback analyzes the past and feed-forward strives to envision and realize a desired future.

These differences are summarized in Table 6.1.

REFERENCES

Brandenburger, A.M., and Nalebuff, B.J. (1996). *Co-opetition*. New York: Doubleday.

Camillus, J. (2015). Feed-forward systems: Managing a future filled with wicked problems. *Rotman Magazine*, (Winter 2015), 52–59.

Chakravarthy, B., and Lorange, P. (2008). *Profit or Growth? Why You Don't Have to Choose*. Upper Saddle River, NJ: Pearson Prentice Hall.

Courtney, H., Kirkland, J., and Viguerie, P. (1997). Strategy under uncertainty. *Harvard Business Review*, 75(6), 67–79.

Dong-Gil Ko, D., and Camillus, J.C. (2001). Managing the future: Planning paradigms and scenario development. *General Management Review*, 3(1): 21–31.

Eisenhardt, K.M., and Tabrizi, B.N. (1995). Accelerating adaptive processes: Product innovation in the global computer industry. *Administrative Science Quarterly*, 40(1): 84–110.

Hart, S.L., and Christensen, C.M. (2002). The great leap: Driving innovation from the base of the pyramid. *Sloan Management Review*, 44(1): 51–56.

Kolko, J. (2015). Design thinking comes of age. *Harvard Business Review*, 93(9): 66–71.

Lindblom, C.E. (1959). The science of "muddling through". *Public Administration Review*, 19(2): 79–88.

Porter, M.E. (1980). *Competitive Strategy*. New York: Free Press.

Prahalad, C.K., and Hamel, G. (1990). The core competency of the corporation. *Harvard Business Review*, 68(3): 79–91.

Quinn, J.B. (1980). *Strategies for Change: Logical Incrementalism*. Homewood, IL: Irwin.

Trigeorgis, L. (1996). *Real Options: Managerial Flexibility and Strategy in Resource Allocation*. Cambridge, MA: MIT Press.

Veliyath, R. (1985). Feedforward Orientation in the Strategic Management Process: A Contingent Choice (Doctoral dissertation, University of Pittsburgh).

7

Frugal Engineering and Innovation

The third construct drawn from the responses to the challenges giving rise to the need for Business of Humanity (BoH) Strategies is that of frugal engineering. This is not a topic that is commonly addressed in the arena of strategy, being seen as more logically in the purview of industrial engineering or operations management. It is a critically important construct, however, in the context of BoH Strategies. Disruptive innovation is both a prerequisite and a consequence of addressing the needs of the 4 billion people at the bottom of the pyramid. Given the limited, to say the least, disposable income and purchasing capability of customers at the base of the pyramid, such disruptive innovation has of necessity to be frugal. Meeting the price point pressures that exist in the BRIC economies is the challenge that frugal engineering seeks to meet.

Transnational corporations have been entering new and emerging markets since the early 1990s, especially Brazil, Russia, India, and China. Their track record of success, however, has been somewhat uneven. Disquietingly, there are signs of late that this "emerging market rush may end up like a giant version of the first Internet boom 15 years ago. The broad thrust was right but some big mistakes were made" (*The Economist* 2014). One of the big mistakes, one might argue, is that transnationals sought to "glocalize" existing products, marginally changing them before attempting to sell them in the emerging markets. In stark contrast, domestic start-ups in the emerging economies were engaging in what was known in India as "jugaad"—a term in Hindi that can be broadly translated to somehow get things to work (and perhaps cut corners in the process) with scarce resources.

Jugaad has been avidly seized by management gurus in the developed countries as representing the native genius of Indians and Chinese and the process to be adopted to respond to the price point pressures of the emerging economies. In our estimation, this is also a "big "mistake." Jugaad has the ethos of bricolage and is more suited to one-off applications, like making a windmill using cast-off bicycle parts or, just to drive the point home, a telephone out of tin cans and string. Frugal engineering on the other hand results in (and both these examples are real) notebook computers that cost $35 or functioning smartphones that sell for $5.

Frugal engineering is a structured approach, unlike "jugaad." Jugaad, like bricolage, is an *ad hoc* approach using found objects to create the desired product. It is not readily replicable and often results in jury-rigged and unreliable products. In contrast, frugal engineering is designed to create efficient products at low cost that effectively meet customers' requirements.

There is a widespread lack of awareness of and misapprehensions regarding frugal engineering, Therefore, before we explore the elements of the construct that were identified in Chapter 4, we would like to share our understanding of frugal engineering.

Another word for this construct is frugal innovation—a term coined by the former CEO of Renault-Nissan Carlos Ghosn. Ghosn terms this as a process of "achieving more with fewer resources." He was impressed with the ability of Indian engineers to innovate cost-effectively and sought to harness their expertise in turning out a small car that could rival the world's cheapest car, the Nano, produced and sold by the Tata Group for—amazingly—$2000. He first formed an alliance with the Bajaj group in India for this purpose, but met with limited success. He later transferred Gerard Detourbet to the company's facility in south India to develop this global small car, the Renault kwid.

What are the characteristics of frugal innovation? The first most certainly is simplification. Engineers design and build products and systems. This can include consumer products (razors, footwear, staplers, etc.), engineered products (cars, planes, blenders, etc.), and systems (to reduce paperwork and approvals, alleviate queues at theme parks, robust supply chains for movement of goods, etc.). Since engineers are technologically savvy, they often tend to "over-engineer" and add "bells and whistles" to their designs. Such products are more expensive to build and maintain, thus putting them out of reach of a large segment of population.

Kumar and Puranam* opine that frugally engineered products for the bottom of the pyramid for the Indian market must also be robust,

---

* See: http://nraoetkc.blogspot.com/2013/09/frugal-engineering.html

portable, "defeatured" and not contain unnecessary options, utilize leap-frog technology, and focus on special services. On the other hand, Sehgal et al. (2010) choose a wider scope and identify factors such as a focus on bottom-up innovation and organizational ability. Fostering cross-functional teams, top-down support, and nontraditional supply chains provide support to frugal engineering.

Companies like GE did something that many other transnationals were not even attempting. According to Paul Polak and Mal Warwick in their recent book, *The Business Solution to Poverty*, transnationals typically enter emerging markets and find, to their surprise, that consumers are quite unlike those in the advanced countries. They make adjustments to their existing product lineup, using cheaper materials, eliminating features, lowering quality, and rebranding them more cheaply—only to discover that what they want to sell does not meet the needs, expectations, and aspirations of poor customers. In contrast, GE focuses on *empathetically understanding the needs* of not-so-rich customers and works backward to create products that efficiently target these needs.

The competition to the Nano typically went about stripping a car down to its essentials to see if it could then be assembled at the lowest price point. Or even adding a fourth wheel to a three-wheeler! Japan's Suzuki Motor Corporation, the leader in India's car market, which till then produced the cheapest car, the Maruti 800, openly scoffed at Tata's efforts, stating that there could be nothing cheaper than what it was already manufacturing. In the event, they were proved totally wrong. The Nano was far from a stripped-down version of any other small car in the world. Simply put, it was a breakthrough in frugal engineering and design.

Frugal innovation can be defined as the process of exactly matching the needs of a given customer segment with the level of design and complexity in a product, process, or service, and by doing so provide it to the customer at an affordable price, thereby contributing enduring value. The biggest misnomer about frugal innovation is that it is another word for "jugaad," which, as we suggested earlier, can be broadly translated to somehow get things to work (and perhaps cut corners in the process) with scarce resources.

"*Creating Change on a Shoestring Budget*" is the subtitle of a book by Charles Leadbeater (2014) titled *The Frugal Innovator* that succinctly captures the jugaad mindset behind typical attempts at frugal innovation. Leadbeater has been much impressed by what he saw in developing countries where innovations were taking place amid serious resource constraints. The favelas of the Brazilian city of Curitiba thus have a self-sustaining system for recycling household waste. Or in the suburbs of Mexico where a $5 a month healthcare system for mobile phone users called MedicalHome is thriving. There are many such examples of resource-constrained innovations in India and China, such as the Maruta, which is typically a diesel generator strapped onto a quadricycle with the shaft of an old jeep and steering wheel taken from some abandoned vehicle!

According to Leadbeater, such jugaad-type initiatives combine four conditions for frugal innovation—they are Lean, simple, clean, and social. Lean is something we agree with, and social sensitivity is important. Innovations like the Tata Nano have to embody lean engineering principles to become a reality. But there is nothing simple or clean or necessarily social about any of GE's innovations or the Nano for that matter. It can hardly be described as a cannibalized version of a simple four-wheeler that was produced on a shoestring budget by a social process! Simply stated, a jugaad mindset could not have produced a Nano. The Nano represents, in fact, the highest achievement of frugal engineering that not only has a Lean component but also incorporates the best principles of value and quality engineering. The four conditions we affirm, therefore, are Lean, value, quality, and, above all, affordability.

Frugal engineering can effectively support BoH Strategies. Frugal innovation and the engineering that makes it possible is not a knockoff of a more expensive product. It is not low-tech either. Although it has been evolved in the specific context of developing markets—especially addressing the requirements of the bottom of the pyramid—it is perfectly suitable for the developed markets as well. Frugal innovation, in fact, establishes a new paradigm by providing world-class products to a targeted low-income, customer segment of the proverbial socioeconomic "pyramid."

While frugal innovation refers to products designed for the bottom of the pyramid and for sale largely in developing countries, this fascinating concept can be applied at many different levels of the same pyramid. In fact, a frugal engineering index can loosely be defined as "value/cost." In implanting frugal engineering in a company's business model, the intent is to increase value and reduce cost so that a customer can receive a product or service of the high value and quality at the lowest possible price. This comes about thanks to a complete rethink within the concerned organization regarding processes in pushing the limits of the production possibility frontier.

THE FRUGAL CONSUMER

Let us start with the psychology of the frugal consumer in the BRIC economies. A large component of consumer behavior is based on inherent individual values. In shopping for an automobile, a thrifty or frugal consumer will be loath to pay for frills or options that she does not need or perceives of as low value. Hence, manual stick-shift transmissions are found in the large majority of cars sold in India while it is rare to find a car without an automatic transmission in the United States. While cars with automatic transmissions are easier to drive, they consume more gasoline and are more expensive to manufacture and maintain.

The basic model of the Tata Nano, an automobile that has established a unique market segment as a high-quality and inexpensive mode of transportation, does not even have a radio! The segment that it was targeted at was a family of five perched precariously on a two-wheeler in small towns and the Indian countryside. To draw such customers to a four-wheeler, it was important to provide a model that offered a high degree of value at the lowest possible cost. Because of their low incomes, these customers are highly value conscious and shop relentlessly for products that increase their individual value/cost index.

Frugal innovation thus appeals to the emerging value-conscious consumer. The realization that "helping" a consumer and the community is important is not new, but it takes organizational discipline and a sharp focus on frugal engineering to make a profit by doing the right thing. The then chairman of the Tata group, Ratan Tata, told his engineers to design a car that costs $2000. That became a stretch target for engineers like Girish Wagh and his *500-strong super-team* at Tata Motors, motivating reexamination of every automobile engineering concept to radically cut costs and weight to design and manufacture this car at the targeted, incredibly low, price point set by Tata.

The decision to work backward to develop a world-class product at a low price point is common to both GE and the Tatas. As explained by Professor Ashok Jhunjhunwala of IIT Madras, the imperative driving the emergence of disruptive technologies in the Indian context is that the viable price point is extremely low. In India, for example, the income of a majority of the population is less than $200 per month. In the rural areas, a substantial proportion of the population subsists on $60 per month. The *viable price point for products and services to such segments of the country is a BHAG* for decision makers from the developed economies.

Radical cost cutting is integral to frugal engineering. Globally integrated enterprises like Bharti Airtel remain profitable despite rock-bottom tariffs, thanks to innovations like its Minutes Factory, which enables it to expand the subscriber base of its prepaid customers. Another radical cost-cutting measure is that it has outsourced network planning and the IT backbone and converted its fixed costs to variable costs. This has allowed it to invest a minimum amount to set up a network that can handle a threshold level of calls and then wait for the usage to build to spend more. Bharti Airtel's thin margins both require and enable building the traffic volume as the key to profitability.

Another way to increase the value/cost index, according to the *Economist*, which we have cited earlier, is to apply mass production techniques in new and unexpected areas like healthcare. There is no way that Dr. Devi Shetty, the charismatic heart surgeon who heads Narayana Hrudalaya in Bengaluru, and his team could perform the largest number of heart operations in Asia without achieving economies of scale and specialization. His hospital charges $2000 for an open heart surgery which in the West would easily cost $20,000 to $100,000. He has helped start a health insurance scheme, Yeshaswini, through which hundreds of thousands of farmers in the state have access to healthcare.

While scalability is integral to frugal innovation, however, there are other constraints in delivering affordable medical care to the bottom of the pyramid. In 2007, nonprofit D-Rev (or "Design Revolution") was founded with the objective of designing and delivering medical products to people living on less than $4 a day: products such as a ReMotion high performance knee—that goes beyond the jugaad innovation known as the Jaipur Foot—is available at a price point of $80, as against a comparable knee priced at $6000 in developed nations. With the product in hand, the challenge becomes reaching prosthetic clinics around the world.

According to D-Rev CEO Krista Donaldson, although the company decided to manage manufacturing and distribution, the complexity of the market for the D-Rev knee is challenging. There are a limited number of clinics—around 300—in most countries. Some countries have no more than one or two clinics in total. "One of the things that we've seen from our understanding of the number of clinics, but also the number of amputees who need knees, is that once we hit about 70,000 amputees served per year, we hit a new hurdle. And that's the number of skilled prosthetists who can fit the knees," she told McKinsey Publishing's senior MD Rik Kirkland.

D-Rev hopes to get around this barrier by working with prosthetics colleges to help with curriculum development around polycentric knees. Some of the lessons she has learned—not just from D-Rev, but her life before, working in Iraq and working in Kenya—is that when "we talk about innovation and disruption, we're often very focused on the technology. But really, it's the system that is often causing the disruption." Innovation focuses on the physical thing. But how can we innovate the distribution channels? How can we improve some of these less sexy aspects of the product life cycle and delivery of the product affordable to those living below $4 a day?"*

Back to frugal innovation, it targets the emerging value-conscious consumer. The intent to respond to and help a consumer and the community is not enough—it takes organizational discipline and a sharp focus on frugal engineering to make a profit by doing the right thing. When the chairman of the Tata group, Ratan Tata, told his engineers to design a car that should cost $2000, it became the starting point and the driving force to rework the entire value chain. The price point target focused and guided the zero-based, design and engineering efforts of a task force of 500 engineers.

The decision to work backward to develop a world-class product at a low price point is common to successful companies such as GE and the Tatas. As explained by Professor Ashok Jhunjhunwala of IIT Madras, disruptive technologies emerge in contexts like India because of the price point pressures,[†] which serve as a motivating BHAG. In India where, as we mentioned earlier, a substantial proportion of the population subsists on $60 per month, the price point for providing products and services to such segments of the country is almost unimaginably low.

To *empathetically understand the frugal consumer*, digging deeper, beyond the obvious is of critical importance. That sounds like a gratuitous statement. But companies are often surprised by how their customers behave and what they value. Companies assume that they know their customers through customer service or sales. In product development, to gain a real advantage, the willingness to dig deeper can make a huge difference.

---

* See: http://www.mckinsey.com/industries/social-sector/our-insights/designing-for-social-impact-the-d-rev-story
† See: http://www.katz.pitt.edu/boh/case-studies/bttm-pyramid.php

In the case of the frugal consumer, identifying real needs and perceived value is especially difficult and critical. In order to do so, experts like Tom Kublius of Bally Design suggest a four-step process:

1. **Get Out There.** There is no substitute for first-hand observation. That means you have to go where they live. If you serve operating rooms, you should be watching surgeries. If your product is going to be used by villagers, spend a week in a village. For example, while the Tata Nano is hailed as a world class, frugally engineered product, it does not come even with a radio since it was perceived as having little value in a rural area.

2. **Draw Reality.** When you immerse yourself in a new reality, it is important to capture as much of your environment, especially if it's not your own. That can mean taking video, pictures, sound recording, and any other media that will give you a more permanent record. Capturing what you see on paper will force you to create a mental model of the situation. Before establishing Goal Zero, a company that produces solar-powered electricity, which is detailed later in this chapter, Robert Workman took detailed notes in Congo that helped define the specification for his products utilized in Congo.

3. **Learn the Rules.** When Apple chose to make the iPhone accessible to all segments of society and also moved aggressively toward increasing penetration in emerging markets, they quickly realized two challenges: (a) that the iPhone was out of reach for many segments of the socioeconomic pyramid, and (b) customers in India and China wanted the same iPhone with largely the same features but at the lower cost. The result—the iPhone 5s with many bells and whistles and the iPhone 5c that could be customized, looked as good, but cost significantly less.

4. **Prototype Early and Often.** You can never build too early. In developing the Yeti series of solar-powered products, Goal Zero engineers initially created a workable prototype that turned out to be twice the size of a marketable product. After the CEO, Workman, challenged his engineers to do better, they developed a second version that was half the size of the first at a tenth of the cost! *Prototyping and experimentation were found to be essential to meet the price point BHAG.*

FRUGAL ENGINEERING FOR THE BoH

Frugal engineering integrates a wide range of tools, drawn from established bodies of knowledge. Frugality entails the aggressive application of the principles of value analysis, Lean engineering, quality management, and determination of customer needs. The graphic model in Figure 7.1 represents the domain of each of these bodies of knowledge. The intersection of these four domains represents their aggressive application. The Tata Nano clearly has become one of the classic applications of frugal engineering as it has successfully integrated these principles to ensure a high-quality product at the lowest-ever price point, the BHAG of $2000.

According to James Bolton, of Whirlpool Corporation and president of the Society for Value Engineers, *value engineering* is a method developed in the late 1940s by Lawrence D. Miles, an employee of GE. It is a scalable method to evaluate products, processes, and projects from a purely functional perspective. This powerful tool has been applied to all types of products, processes, and projects in manufacturing, construction, government, transportation, healthcare, environmental, and other fields. It is a powerful tool that helps design products and processes that function without any unnecessary bells and whistles that do not provide value to the consumer.

Value engineering works well when all stakeholders are represented in the deliberations. This ensures that the value/cost index of a product or

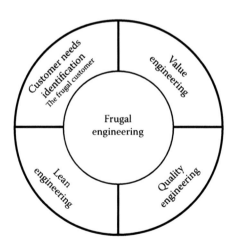

FIGURE 7.1
Circle of disruptive frugality: Four quadrants of frugal engineering.

service is maximized because of the sharp focus on function—a key ingredient in the purchasing decision of frugal consumers who seek value in products. Most value engineering results are a complete surprise to the people that created them! It must be emphasized here that value engineering is not cost reduction that results in eventually cheapening the product; rather, it entails new and different ways of performing the functions demanded by the market place.

Lean engineering helps the manufacturer eliminate waste, thereby reducing cost (and often increasing quality since it reduces clutter and a lack of organization in the workplace). It allows the frugal consumer to receive a higher quality product at a lower cost, again increasing the value/cost index of product or service. The focus of this discipline is the elimination of each element of waste in the manufacture of a product or a service. A novel application of Lean engineering is India's Flipkart's service where books are delivered by a messenger, eliminating the inertia of the Indian Postal Service.

Toyota and Amazon are classic examples of Lean engineering. In Toyota, Taiichi Ohno is considered to be the father of the Toyota Production System (TPS). At the heart of the TPS is a focus on reducing (or eliminating) the three major shortcomings of assembly line work—muda (waste), mura (imbalance or inconsistency), and muri (overburden). Lean focuses on the elimination of muda (waste). There are Seven Wastes—Overproduction, Transportation, Overprocessing, Waiting, Scrap and Rework, Inventory, and Operator Motion. Any work that we do, whether it is in a factory or hospital or office or warehouse, requires us to execute a pre-established process or processes.

Any process consists of a sequence of steps. Each step can be value-adding (VA), non-value-adding (NVA), or necessary-but-not-value-adding (NNVA). From a customer's viewpoint, "value" is a feature (or function) of a product that he or she wants the product to possess. It is the NVA steps that consume and waste the many resources (money, labor, material, equipment, space, energy, information, management decisions, etc.) needed to make any product. Resource consumption increases the price that the customer has to pay to receive the product they ordered. Lean pursues the elimination (or reduction) of waste to satisfy customers by delivering the quality they desire at an acceptable cost.

In the words of Steven Fetch, Director of Global Quality, 3D Systems Inc., *quality engineering* is a field that utilizes common sense and engineering principles combined with statistical methods to improve performance.

Often performance and quality engineering are thought of statically and quantitatively (good vs. bad) for parts or services. However, this approach is a gross misconception because the reality of quality engineering is, in fact, a relentless focus on continuous improvement. This quest for continuous improvement is a never-ending goal to improve systems, processes, and products to achieve higher levels of performance.

While many consider high quality to be achieved by a focus on making good parts, the practicing quality engineer has a focus on the reduction of waste and increasing value to the frugal consumer. The old phrase that high quality costs less most often holds true when the total cost of activities associated with poor quality are understood and calculated. For example, if a product fails in service and needs repair, not only does the company lose goodwill, but also incurs a significant additional cost, thereby wiping out profit margins. The difference between cheap and inexpensive is a crucial one. Today's consumers focus on products with the highest value/cost index.

These disciplines or bodies of knowledge described above must be integrated to design and produce frugally engineered products and services to cater to customer needs at targeted levels of the socioeconomic pyramid. Value engineering is typically applied at the design stage. Lean engineering has broad application in manufacturing, while quality engineering principles can be applied to both stages.

Frugally engineered products can, of course, be produced by creative genius and inspiration, if not serendipity. However, in order to systematically and consistently develop products and services for a frugal consumer, it is imperative that all four disciplines in Figure 7.1 be applied at the different stages of the cradle-to-grave product life cycle. For example, a frugally designed and value-engineered product attuned to customer needs can fail to serve the consumer if it is poorly manufactured without the effective application of Lean and quality engineering. This results in a product or service that breaks down frequently and cannot withstand the rigor of harsh operating conditions that a frugal consumer will count on.

FRUGAL INNOVATION—GOAL ZERO

Robert Workman, the founder of Goal Zero, describes frugal innovation from a perspective that is aligned with the BoH: "We are not a business looking for a humanitarian mission, we are a humanitarian mission wanting to benefit humanity and looking for a sustainable mission." Goal Zero Inc. was named as one of the 10 fastest growing companies in the United States by *Inc.* magazine last year. Seven years ago, Workman was a wealthy, semi-retired entrepreneur. He had just sold his company and not only wanted to travel the world, but also affect it. He chose to visit the Democratic Republic of Congo and initiated multiple business ventures in water, brick building, and trucking industries.

Each time he handed over the businesses to the Congolese, they failed. Workman was frustrated, but instead of giving up, he became even more persistent in his desire to help the expectant villagers. During one of his visits, he realized that the country was blessed with bright sunny weather all year round but, of course, was dark after the sun went down. When he needed a light to work at night, he decided to build a solar-powered light that would power a 100-W bulb and work at night. He bought the parts needed for such a system and was shocked when he realized that it would cost him almost $3000! ($2000 for the 90-W solar panel, $400 for a 100-Ah battery, $500 for a 10-A charge converter, and $10 for a simple light bulb). In a poor country like the Congo, obviously not many people could afford this.

Workman also realized that he could make the most impact, perhaps not by starting new businesses, but by providing DC power to the villagers and allowing entrepreneurs to use this power to build their own businesses. On his return to Utah, he decided to initiate an effort to implement his vision—by building solar-powered electrical generators for use in villages without power. He formed Goal Zero Inc. with a mission "to empower people ... to put power into the hands of every human being" (no pun intended?). He invested $2 million and hired the best young engineering talent he could find and tasked them with developing high-quality, best-in-class, solar-powered generators. He began with seven employees in Salt Lake City, Utah, and his engineers rewarded him with a "black box," 12" by 18" by 18" costing approximately $1000.

But he was not satisfied. Having understood his customers' needs perfectly, he quickly realized that this bulky and expensive box would not be affordable in Congo. He challenged his employees to do better. He personally *sourced the globe* for high-quality, reliable, and robust components that also happened to be inexpensive. He subjected these components to a rigorous quality engineering and testing program. His engineers value engineered the product (named GoBe and later renamed the Yeti 150), a simple LED light powered by a solar panel and a battery. He utilized

Lean engineering to cut manufacturing costs, but maintained quality. Eventually, his engineers designed and produced a new "black box" that was less than 15% of the original size—it was now a much more manageable cube of 8" by 8" by 8"—the application of Lean manufacturing sliced its production cost to $100 or a tenth of the original prototype!

The innovations driven by the needs of the Congo enabled Goal Zero to access developed markets. *The innovations were enabled by and also connected developed and emerging markets.* The Yeti 150 is an extremely popular product that is now sold by all major departmental retailers in the United States including R.E.I., Cabela's, Costco, and Amazon. Its current mode has a capacity of 150 Wh and 14,000 mAh and can be charged by solar power or plugged into a wall outlet. It retails for $199.95 and is currently sold out because of high demand! It is also used by Workman's adopted villages in the Congo. Workman now provides DC power at a nominal cost via the Yeti 150 to a few villagers in the Congo. Local entrepreneurs have used this DC power to set up new and innovative businesses in the villages, ranging from a cell phone charging station to a store selling frozen goat meat, and even a movie theater. The average wages in these villages have skyrocketed from $30 per month to $600 per month. The frugally engineered Yeti 150 services the poorest of the poor and the richest of the rich.

Goal Zero now offers a variety of solar-based products, including power packs, solar panels, solar kits, and off-grid, on-the-go products like chargers, flashlights, and cameras. The company has 110 employees and $30 million in annual sales. Just as importantly, the values of Workman are all-pervasive. When Hurricane Sandy blacked out New York City, Goal Zero employees swiftly began packaging and transporting kits to send to the power hungry area—all without the knowledge of the executives! According to the Deseret News, Workman's reaction was "that I (he) was never so proud of his people. It wasn't a top-down operation. It was bottom-up. It was touching to see these young people step forward and do what they did, on their own." Goal Zero has also helped victims of the Haiti earthquake, Japan's tsunami, and Typhoon Haiyan.

FRUGAL INNOVATION—THE INDIAN MARS MISSION

India cemented its position as a global leader in disruptive frugal innovation with the launch of its Mars Mission named the Mangalyaan (or

"Mars Craft" in Hindi) on November 2013 in a small island in the Indian Ocean. Previous space missions had shown how these forays into space have had far-ranging direct and indirect benefits to the common man including the development of early warning systems for the prediction of life-threatening hurricanes and storm systems, the identification of optimal fishing areas, the estimation of water levels in underground aquifers, and so on. It is expected that the Mangalyaan will yield similar benefits over the next decade.

The Mangalyaan mission confirmed that frugal engineering can be applied to the most sophisticated levels of technology. The Indian Space Research Organization (ISRO), a Government of India organization that is the equivalent of NASA, was spectacularly successful on multiple dimensions in its Mangalyaan mission as follows:

- The entire project (from approval to launch) was completed between 15 and 18 months, an unusually short period for a mission of such complexity. A Mars mission normally uses a Hohmann Transfer orbit to send a spacecraft to the red planet. This orbit requires the least amount of energy for the journey. But to achieve this orbit, the Mangalyaan spacecraft had to be launched in only a specific period (called the "launch window"), which enables the craft to take a path where the distance between the moon and the earth is at its minimum. When the project was approved by the Indian Government, the closest launch window was 15 months away. A failure to utilize this window meant that the next launch was possible only in 2016. The odds were stacked against ISRO. But ISRO believed in itself and took a calculated risk to go ahead with the seemingly impossible. Its young staff was passionately committed to the project and worked long hours and even a few all-nighters to ensure readiness at launch.
- The cost of the project (approximately $75 million) is even more surprising, especially when compared to NASA's recent MAVEN Mars mission, which costs approximately $671 million.* While many attribute this to the low cost of engineering talent in India (an aerospace engineer is paid approximately a fifth of her equivalent in the United States), this does not tell the entire story and accounts for

---

* See: http://articles.economictimes.indiatimes.com/2014-02-18/news/47451367_1_space-agency
  -indian-space-research-organization-mission

around half the cost differential. The other half of the story is that the ISRO has long instilled a culture of frugal engineering throughout its organization.

ISRO—started in 1969 by the visionary scientist Dr. Vikram Sarabhai—has grown into a pioneer in low-cost space rocket launches. It is also a poster child of the application of frugal engineering in a hi-tech environment.

In order to study the cost differential attributed to frugal engineering, it is useful to first understand ISRO's mission and its understanding of the needs of the primary sponsor (or customer), the Government of India. ISRO is tasked to "promote the development and application of space science and technology for the socioeconomic benefit of the country." In developing the Mangalyaan mission, therefore, they chose to focus on developing a frills-free Mars Orbital Mission at the lowest possible cost but with no compromise in quality.

In designing the spacecraft, systems were value engineered and redundant systems and models were eliminated. The major drivers of the frugal engineering in ISRO were as follows:

- Value engineering: Adapt existing technology as much as possible with a focus on commonization to increase value and reduce costs. The Mangalyaan mission borrowed system platforms generously from previous missions including the gyro system, the attitude control system, the star tracker system, and so on. Honda also follows this technique by utilizing and tracking common parts and systems across its multiple platforms.
- Lean engineering: Unlike similar missions undertaken at other national space agencies, where multiple models are built and tested (e.g., a qualification model, a flight model, and a flight spare), the Mangalyaan mission utilized a single model that had to be built right the first time. The overall project lead time was significantly shortened because of planetary positions and the focus on minimization—creating a BHAG. This reduced all but the essential builds to the actual craft itself, thereby increasing the value cost index and reducing all nonessential non-value-added activities.
- Focus on the mission objective: The Mangalyaan mission had a single objective—to establish a life marker on the Red Planet. It consequently carried only five instruments that helped measure the presence of methane on the planet. In contrast, NASA's Mars

mission, the MAVEN, had a broader objective, to "explore the Red Planet's upper atmosphere, ionosphere and interactions with the sun and solar wind,"* forcing it to carry eight instruments to meet the mission objective. The single focus in a limited mission objective of the Mangalyaan meant that it was simpler to assemble and had a significantly smaller payload than the MAVEN.

ISRO engineers utilized value engineering principles integrated with quality engineering, which allowed them to minimize cost and maximize quality and value. In space engineering, where conditions are tough and costs of failure are high, it is not easy to adapt technology. It also involves more risk, but ISRO was willing to take—and manage—that risk. "Traditional ways take a long time, and for ISRO, time was of essence," says Alok Chatterjee, project engineer at Jet Propulsion Laboratory of NASA. "So, its approach is an innovative way to do space missions."

Terri Bresenhem, from GE's Health Care Division, sums up ISRO's approach nicely: "If necessity is the mother of invention, constraint is the mother of frugal innovation" (*New York Times* 2014). Mangalyaan entered Mars' orbit on September 24, 2014. The breaking of barriers in the field of science and technology is a huge step toward the development of a country. ISRO's brand of frugal engineering certainly helped make a huge contribution not only to India's development but also to raise the self-esteem and pride of the engineering establishment in the country. The economic value of this is difficult to estimate but is certain to be of very great value.

The credibility of the Indian space launch capability was enhanced by Mangalyaan. This added credibility helped strengthen the Indian commercial space launch business—attracting *customers from the most advanced economies and leading edge companies.*

Each of the elements of the "frugal engineering" construct, developed in Chapter 4, was supported by the examples described above. These elements are as follows:

- Relate empathetically to the customer
- Set price point–related BHAGs
- Employ skunk works and super-teams
- Nurture and connect innovation ecosystems

---

* See: http://www.nasa.gov/mission_pages/maven/overview/index.html

- • Cross-industry connections
- • Cross-markets/countries connection
- Rapid prototyping and experimentation

In light of the preceding discussion, the elements of the construct take on practical meaning and significance, and offer operational guidelines. The relevance and importance of the frugal engineering as a construct both derive from and support fundamental and generalizable characteristics of BoH Strategies, namely, global reach, serving low-income segments, price point BHAGs, and reverse innovation.

REFERENCES

"Emerge, splurge, purge", *The Economist*, March 8, 2014.
"From India, Proof that a Trip to Mars doesn't have to break the bank", *New York Times* (Saritha Rai), February 17, 2014.
Leadbeater, C. (2014). *The Frugal Innovator: Creating Change on a Shoestring Budget.* London, UK: Palgrave.
Sehgal, V., Dehoff, K., and Panneer, G. (2010). The importance of frugal engineering. *Strategy+Business*, Issue 59, Summer 2010.

8

Forging Business of Humanity Strategies

In this chapter, we will demonstrate how the three constructs—Identity, Feed-forward Processes, and Frugal Engineering—in the Business of Humanity (BoH) Management Framework interact and reinforce one another to support the formation and implementation of BoH Strategies. For ease of reference, the BoH Management Framework is reproduced in Figure 8.1.

As we discussed in Chapter 5, Identity serves as the anchor, beacon, and compass for BoH Strategies. The three components of Identity—values, aspirations, and competencies—presented in Chapter 5 support these functions of Identity. The organization's values serve as the anchor that provides constancy and focus for BoH Strategies. Aspirations are the beacon, the goal of BoH Strategies; they provide the framework for articulating the organization's vision. Competencies, existing and needed to realize the organization's vision, determine the feasible strategic alternatives and initiatives that should constitute the organization's BoH Strategies.

The process of developing BoH Strategies starts with the dynamics within each of the three constructs of Identity, Feed-forward Processes, and Frugal Engineering. While the internal dynamics of Feed-forward Processes and Frugal Engineering have been discussed in Chapters 6 and 7, the critical importance of Identity requires that we examine the internal dynamics of this construct in some more detail here.

In shaping the construct of Identity, the first step is to surface and formally articulate the values of the organization. Surfacing the values of the organization can be an informal exercise deriving from a conversation within the top management team of the organization. Alternatively, a more formal process engaging a wide spectrum of personnel within the organization can be adopted. Vodafone CZ, for instance, engaged task forces of employees

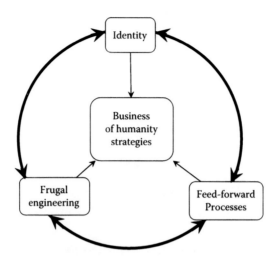

FIGURE 8.1
The BoH Management Framework.

to help define the existing values and how they should evolve. Vodafone CZ's then CEO, Muriel Anton, was intrigued by the evocative notion that emerged from the task forces. The task forces came to the conclusion that Vodafone CZ at that time possessed the personality and values of a creative, 19-year-old person. With its growing emphasis on business customers, in addition to its initial focus on individual customers who appreciated the 19-year-old persona, the task forces suggested that Vodafone CZ adopt and seek to assume the values of a person in his or her late 20s or early 30s. This meant a greater emphasis on values such as reliability and tenacity. Of course, the task forces gave undisputed primacy to the value espoused in their long-standing policy: "Break All the Rules for the Customer."

The "humane" dimensions that characterize BoH Strategies, which were identified in Chapter 1, can serve as an inventory of possibilities for values that can be espoused by the organization. To recapitulate, some of the possibilities are safety, quality, environmental sustainability, diversity, gender equality, and integrity.

These values inform and color the aspirations—goals and vision—that are a part of the organization's Identity. For instance, a value system that commits to environmental sustainability can lead to the aspiration of a textile manufacturer such as Arvind to become the leading producer of organic denim. While many in the organization may contribute to the definition of its aspirations, the final determination of the organization's aspirations is very much the purview of the top management team. This recommendation stems from one of our favorite dicta, which is Cyrus the Great's maxim: "Diversity in Counsel but Unity in Command."

In the context of BoH Strategies, it is important to remember that these aspirations should possess the character of BHAGs or stretch goals. The BHAGs and stretch goals drive innovation and Frugal Engineering, which are of the essence in BoH Strategies.

Finally, identifying the organization's competencies is a task that needs to be carried out with objectivity and insight. Objectivity is difficult to achieve but easy to understand. The criteria that are widely employed for identifying and evaluating competencies (Barney and Hesterly 2005) and that might help in gaining objectivity include the following:

- Valuable—that the competency has the potential to add economic value through building competitive advantage
- Appropriable—that the competency can be owned or possessed by the organization
- Rarity—that it is not widely available
- Inimitable—that other organizations, especially competitors, cannot readily replicate the competency
- Organizational—that the competence is compatible with the organization's characteristics and existing capabilities

Insightfulness is both difficult to achieve and to understand. Perhaps an example may help. The first president of FedEx, Art Bass, when looking at the competencies that FedEx possessed and needed, discussed the competencies possessed by the major car rental companies such as Hertz, Avis, and National. Objectively, one could argue that the car rental companies had capabilities built on their multiple locations and their proximity to major airports—capabilities needed and possessed by FedEx. Art Bass's insightful analysis, however, led to the further understanding that car rental companies displayed the rare ability to interact effectively with retail, individual customers, as well as with corporate customers, a capability also needed by FedEx.

LINKING IDENTITY WITH FEED-FORWARD PROCESSES

The definition of the organization's Identity arising out of the processes described above directly interact with Feed-forward Processes. Feed-forward Processes, through the development of possibility scenarios and transformational scenarios, lead to the identification of ranges of robust actions and enablers. Evaluating the relative importance of these robust actions and enablers demands a set of criteria and decision processes that can cope with both quantitative and qualitative criteria. The Identity construct provides the criteria that can be employed, in addition of course to the traditional profit projections, and return on investment (ROI) and discounted cash flow (DCF) calculations.

The values that are part of the Identity construct, if explicitly stated, readily enable the recognition of alternatives that are not aligned with the organization's policy profile. Of course, they also enable the identification of alternatives that support the value system and related policies of the organization. This assessment does not require a formal decision process beyond a conversation among executives of the organization.

Assessing the alternatives through the lenses of aspirations and competencies however requires a structured decision process, as the complexities that are involved as well as the need to integrate the outcomes along multiple dimensions cannot be dealt with through conversations. However, aspirations and competencies are qualitative considerations that do not lend themselves to typical optimization processes. There are, fortunately, decision models that are well suited to address the qualitative, complex, multidimensional characteristics of criteria such as aspirations and competencies.

We have found two approaches to linking Identity and Feed-forward Processes to be effective:

1. Creating a template or profile for assessing the fit between strategies developed by Feed-forward Processes and the elements of Identity. This technique is labeled "**Dimensions of Strategic Choice.**"
2. Employing **multi-criteria decision models** that employ values, aspirations, and competencies as the criteria with which to evaluate strategic initiatives.

Dimensions of Strategic Choice

This approach to linking Identity with Feed-forward Processes is a visual device that received an award from the Foundation for Administrative Research as the most significant contribution to "corporate and organizational planning" in 1983. Despite being developed three decades ago, it continues to be a useful approach. An example of the application of this methodology is perhaps the best way to communicate it.

Consider the example of an organization in the health arena, which opted to remain anonymous in a case study written about it, in which it adopted the fictitious name Metrohealth (Camillus 1996). When engaged in a strategic planning exercise, the top management and the Board found themselves faced with a values-laden choice. The choice was whether or not to provide healthcare to patients who did not have the personal resources or insurance to pay for the service. This is a choice that most health organizations in the United States face, but often sidestep, or try to ignore.

When analyzing the issue, the organization realized that this choice would affect other important decisions that defined its strategic positioning or profile. The four other critical decisions that would be largely determined by this fundamental choice of "type of client" to be served included the following:

- Revenue sources—to be tapped by Metrohealth, either in the form of insurance companies or a greater reliance on philanthropic sources for support
- Diversification posture—to aggressively seek diversification opportunities into new arenas or to be satisfied with the scope of the current business
- Technology posture—to develop or adopt leading-edge technologies or employ technology that is proven and affordable
- Growth orientation—to choose to emphasize rapid growth or resolve to grow to the extent that philanthropic support allows

The Board was split on the issue of the "type of client" to be served, with the business executives on the Board tending to favor one position, and the medical professionals, clergy, and social workers on the Board tending to take the opposite view. Ultimately, it was difficult for the Board to choose formally not to provide healthcare to anyone who needed it, because most of them were of the view that healthcare is a fundamental human right. The consequences of this fundamental, values-driven choice are indicated by the arrows connecting the dimensions of strategic choice in Figure 8.2.

Examples of both types of health organizations abound—that is, those choosing to serve anybody needing healthcare and those choosing to focus only on patients with the ability to pay for the care they need. It appears logical to presume that focusing on the latter may better ensure the organization's continued economic viability and independence. However, from a BoH perspective, the former choice, serving anybody needing healthcare,

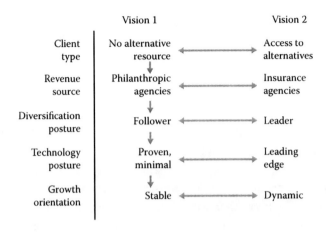

FIGURE 8.2
Dimensions of strategic choice: Metrohealth example.

would be the preferred alternative. Interestingly, there are powerful examples of the viability of such a BoH-driven choice. Perhaps the best known example is Aravind Eye Hospital,* which is a brilliantly successful, rapidly growing organization based in India, which started with the mission of eliminating blindness worldwide, with a commitment to not charging a fee to clients but leaving payment entirely to their discretion. The integrity of their commitment was evidenced by their practice of even arranging free transportation and sustenance to indigent clients from rural areas without the means of paying for travel to the organization's sites; and the quality of care and outcomes offered by Aravind significantly exceeded the levels achieved in developed countries. Other examples of healthcare organizations adopting a BoH stance include Boston Medical Center and St. Jude Children's Hospital in the United States.

Creating a profile along key strategic dimensions can serve as a screen or template against which to test the actions and programs generated by Feed-forward Processes. If they violate the values the firm has chosen to espouse, they need to be discarded or modified.

The Dimensions of Strategic Choice approach illustrated above is more powerful than the popular "strategy canvas" approach to profiling strategy. An example of a strategy canvas for the business school discussed as an example of visioning in Chapter 6 is presented in Figure 8.3. While the strategy canvas approach may seem similar in appearance to the Dimensions of Strategic Choice, which is the approach advocated here, it

---

* See: https://www.youtube.com/watch?v=65TLzg7GShw

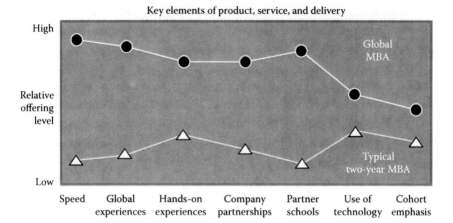

Key elements of product, service, and delivery

FIGURE 8.3
Strategy canvas: Business school example.

does not explicitly assess the logical or practical consistency of the organization's choices across the various dimensions.

Multi-Criteria Decision Models

Approaches that are more structured can be adopted, employing multi-criteria decision models to link the actions and programs generated by Feed-forward Processes with the values, aspirations, and competencies defined in the firm's Identity. Strategic initiatives, identified by Feed-forward Processes through possibility scenarios and value chain analysis, can be prioritized by employing certain of these decision models. The values, aspirations, and competencies that form Identity would serve as the criteria employed in these decision models. For these decision models, qualitative criteria and judgments are perfectly functional and acceptable. Even if the strategic initiatives are extremely different in character—for example, one being technology development to reduce the firm's carbon footprint and another being a risky start-up in a BRIC country—the decision tool identified below has been proven to work.

There is a widely employed decision tool based on a methodology called the Analytic Hierarchy Process (AHP) that responds well to the complex characteristics of these choices. It has been embraced by the management of major corporations such as IBM and 3M, agencies such as NASA and HUD, and by governments of major countries such as China and Indonesia.

The process employed by the AHP methodology is also intrinsically valuable. The process helps both to share and to inform the understanding and priorities of participants in the decision process.

Decision software packages (e.g., Expert Choice and Super Decisions) utilizing variations of this methodology are available from different vendors. A package incorporating this tool, which offers a free version, is available on the web.* These software packages make the use of this powerful tool very simple. The software enables participants in the decision-making exercise to engage in a process of sharing their views to arrive at a consensus understanding of the relative weight to be given to criteria such as the firm's values, aspirations, and competencies. The weights are derived from making a sequence of simple comparisons between two measures at a time, regarding how important each is to making progress toward the overall goal or vision of the organization.

The AHP software allows for the weighted criteria to be applied to alternatives and strategic initiatives. The relative contribution that each strategic initiative makes toward realizing the vision is calculated by the software. Importantly, the software programs enable sensitivity analyses and implicitly offer guidance about how the initiatives can be redesigned to increase their effectiveness.

A hypothetical matrix generated as the outcome of this decision technology is illustrated in Table 8.1.

The software programs offer the power of sensitivity analyses that instantly enable changing the relative weights of the criteria, and the ability to visually comprehend how robust the priorities of the alternatives are. The software is powerful and flexible enough to respond readily to changes in the environment and to new information gained when implementing the strategic initiatives.

The Advantages of Aligning Feed-Forward Processes with Identity

Feed-forward needs the stimulation and the guidance that Identity can provide. In a world of uncertainty and complexity, Feed-forward Processes need an anchor to provide a sense of self and place, a beacon to provide direction, and also a chart to point out the rocks, shoals, and dangerous currents along the way. The organizational Identity provides

---

* http://www.superdecisions.com

TABLE 8.1

Illustrative Priority Matrix of Strategic Initiatives: Relative Contribution of Initiatives to the Goal of "Profitable Progress Toward Becoming a Socially Responsible Global Technology Leader in Selected Sectors"

Criteria / Initiatives	Cash Flow Impact (Weight: 0.26)	Sustainability Index (Weight: 0.15)	H.R. Capability (Weight: 0.12)	Technology Enhancement (Weight: 0.25)	Growth (Weight: 0.22)	Overall Score (Max: 1.00)
Invest and grow in BRIC markets	High	Medium	Strong	Significant	High	0.98 *Rank 1*
Subcontract to minority-owned firms	Marginal	Medium	Strong	Negligible	Medium	0.46 *Rank 6*
Reduce carbon emissions from older plants	Negative	High	Limited	Moderate	Low	0.67 *Rank 5*
Accelerate shift to green energy	High	High	Limited	Significant	Medium	0.84 *Rank 3*
Acquire South African operation	High	High	Limited	Negligible	High	0.87 *Rank 2*
Invest in process improvement research	Moderate	High	Strong	Significant	Medium	0.75 *Rank 4*
Revise service agreements	Moderate	Low	Poor	Negligible	Low	0.37 *Rank 7*

all three. Values identify the boundaries of acceptable strategic initiatives and provide constancy though disruptions. Aspirations are the beacon that motivate action and stimulate creativity. Competencies are the compass that determines the route to take. Either or both methods—the dimensions of strategic choice template or the multi-criteria decision models—of linking Identity to Feed-forward Processes will provide valuable and useful guidance.

While we have been looking at alignment as a unidirectional phenomenon from Identity to Feed-forward Processes, it can also be a two-way street. Feed-forward analyses may come up with possibilities that lead to firms changing their aspirations. New possibilities may motivate the development of new competencies, and, in extreme cases, it is possible to conceive of a future that may warrant a change in the firm's value system. For instance, values that are based on religious tenets may have to give way to more humanistic values as social mores, government regulations, and laws change. Whatever the directionality, effective BoH Strategies are built on the symbiotic relationship between these two constructs.

LINKING FEED-FORWARD PROCESSES AND FRUGAL ENGINEERING

Frugal Engineering and innovation are the keys to implementing the strategic initiatives emerging from the interaction of Identity and Feed-forward Processes. GE's medical systems business unit's development of ECG and ultrasound machines in India and China (Immelt et al. 2009) is an inspirational example of how visionary feed-forward and creative frugal engineering work in tandem. The GE example surfaces three dimensions of the linkages between Feed-forward Processes and Frugal Engineering that are to be managed.

The three dimensions of linkage are content, process, and structure. The content dimension defines the focus of Frugal Engineering efforts. GE's focus on fundamental human needs such as health and water, its commitment to being an industrial organization, its strategy of meeting the needs of large emerging markets, and its competency in reverse innovation drove its efforts to frugally engineer medical equipment. In general, the content dimension incorporates the outcome of the interaction between organizational Identity and Feed-forward Processes. The strategic initiatives that receive the highest priority as the outcome of the interaction between organizational Identity and Feed-forward Processes are the focus of Frugal Engineering efforts. Also, the visionary goals that emerge from Feed-forward Processes serve to motivate Frugal Engineering projects.

GE's engagement of personnel, throughout the length and breadth of its organization, in the planning process and in guiding and strengthening the process of implementation of strategic initiatives is intense and continuous. The process dimension of linkage between Feed-forward Processes and Frugal Engineering should similarly engage individuals responsible for visionary planning and creative, out-of-the-box engineering. Engagement in the process would positively affect the content dimension.

Having the individuals responsible for Frugal Engineering share in the process of visionary feed-forward planning offers the possibility of two beneficial outcomes. First, these individuals may suggest possibilities that may not otherwise be considered by those primarily responsible for Feed-forward Processes. Second, fully understanding the significance of realizing the objectives of the strategic initiatives and accepting the importance of BHAGs will further motivate those responsible for Frugal Engineering.

Coming to the third dimension of linkage—that of structure—GE's bold shifting of emphasis on profits from global product-line profits to the profits generated by each country is a great example. Reverse innovation, which is an important element of GE's long-term sustainability, requires investment in innovation and Frugal Engineering in the domain of emerging economies. Such investments are unlikely if global product line profitability is the focus, because the profits of individual product lines in an emerging economy will be on the relatively lower side. By aggregating the profitability of all product lines in an emerging economy, the country manager gains significance in the corridors of power and can influence resource allocation decisions, both financial and human, toward Innovation and Frugal Engineering to meet the needs and price point pressures of customers in local markets.

A structural device that can bring about the integration of Frugal Engineering and Feed-forward Processes is the use of "super-teams" that include individuals who are responsible for both. Such super-teams can be employed when engaging in both Feed-forward Processes and Frugal Engineering. Of course, the process linkage might be such that Feed-forward Processes and Frugal Engineering are parallel activities that continuously interact and such interaction could be effectively supported by super-teams.

LINKING FRUGAL ENGINEERING AND IDENTITY

The linkage between Frugal Engineering and Identity is primarily accomplished through the connection that both of them have with Feed-forward Processes. We have discussed how Identity guides Feed-forward Processes and how Feed-forward Processes drive Frugal Engineering. In addition to this indirect but powerful linkage, it is again useful to think of the three dimensions—content, process, and structure—of possible linkages that can be employed to connect Frugal Engineering with Identity.

In terms of the content linkage, the outcomes of Frugal Engineering will cause Identity to evolve. The competencies and aspirations that are important parts of Identity will be enhanced and informed through the developments and innovation brought about by Frugal Engineering.

The process and structure linkages can both be managed through the use of super-teams, enlarging the membership of the super-teams that integrate Frugal Engineering and Feed-forward Processes to include individuals responsible for Identity.

Linking Identity with Feed-forward Processes and Frugal Engineering creates the framework that constitutes the management system for the creation of BoH Strategies.

REFERENCES

Barney, J.B., and Hesterly, W.S. (2005). *Strategic Management and Competitive Advantage*. London, UK: Pearson Education.

Camillus, J.C. (1996). Reinventing strategic planning. *Strategy and Leadership*, 24(3): 6–12.

Immelt, J., Govindarajan, V., and Trimble, C. (2009). How GE is disrupting itself. *Harvard Business Review*, 87(10): 56–65.

9

Realizing the Promise of Business of Humanity Strategies

Creating and implementing Business of Humanity (BoH) Strategies in an organization require the organization's leadership to be at the point where we hope we have now brought readers of this book firmly believing in the synergy of social benefit and economic value. Beyond the inductive understanding and deductive reasoning that we have offered in support of the BoH Proposition, the organization's leadership needs to be committed to BoH Strategies at the most fundamental level, with this commitment deriving from their core values. Such a fundamental commitment is necessary because BoH Strategies require significant creativity in their formulation and great dedication in their implementation. It is all too easy to achieve short-term profits by disregarding the business's responsibility to society.

Several years ago, one of our clients drove home this point, offering us one of our most valuable learning experiences. This client when engaging us explained, with the uniquely American capability for evocative neologisms, that his portfolio consisted of "schlock" businesses. "Schlock" businesses according to him focused on clientele at two ends of the spectrum of disposable income. These businesses, for low-income customers, include vocational schools, providing advice on how to get rich, gadgets sold to insomniacs, and used cars. For high-income customers, these businesses are typified by fashion items hawked by celebrities, and luxury items whose primary appeal is exclusivity resulting largely from obscenely high prices. A schlock business that spans both ends of the income spectrum is gambling casinos. The reason he used the term "schlock" to define these businesses is that very often these businesses exploit high-income customers and rip off low-income customers.

That our client chose to focus on "schlock" businesses addressing low-income customers and the way he chose to run these businesses were eye-openers. The value proposition that he committed to was to run the businesses with integrity and in a manner that they provided significant benefits to the customers. He was passionate in his belief that not only was this the right thing to do in ethical terms, but it also was the only way to ensure long-term economic sustainability. He reengineered the entire value chain of his businesses with a view to providing value to his customers. For instance, in the vocational schools that he owned, only those individuals who could benefit from the training offered and who were likely to find gainful employment were admitted. Providing value for money was a challenge that he met with great success in economic terms by creating innovative business models that ensured that his customers received significant value for the money they invested. It was an early indication to us of the power as well as the challenging nature of the BoH Proposition.

The choice that this client highlighted for us is one that is faced, perhaps less explicitly, by most businesses. The choice can be made to run the business in a "schlock" fashion or to run it with integrity to provide good value to the customer and to share value with stakeholders. For instance, private equity firms can operate by bringing the best management practices and needed resources to turn around poorly run businesses.

Or they can adopt a schlock strategy focusing on generating cash by drastically cutting costs and selling off parts of the business. Euphemisms such as restructuring and rightsizing are often employed for layoffs that damage the economic sustainability of the business because of the loss of essential tacit knowledge and the inevitable decline of employee morale. The cash thus generated can be used to buy back shares, thus increasing Earnings Per Share (EPS), which in turn often improves the price-earnings ratio because of superficial analyses by investors that overlook the damage to the strategic capability of the business. Selling off the hollowed-out business to take advantage of the increased market value before the performance of the business declines can result in substantial profits for private equity firms that choose to operate in such a schlock fashion. Such profits do not take much skill to generate, though one can argue that there is a limited amount of times in which such an approach can be employed.

The pressure for immediate profits in most economies makes it tempting for managers to choose to employ schlock strategies. The quarterly EPS regime in the United States is an extreme example of a context where enormous pressure for immediate profit increases exists. Recent examples abound. The price gouging behavior of Valeant Pharmaceuticals, which led to its downfall, could perhaps have been anticipated. According to Forbes, Valeant was an example of "rogue financial operators posing as pharma companies."* The Valeant story illustrates a variation of the schlock strategy adopted by private equity firms with debatable core values that we discussed earlier and, as Forbes pointed out, should not have been a surprise. That Valeant, when its behavior was publicized and excoriated, suffered a major drop in value also should not be a surprise.

What may be surprising, and offers a salutary lesson of considerable relevance to us, is the sorry story of companies such as Wells Fargo Bank and Volkswagen. These renowned companies, with long and storied histories, engaged in 2015 and 2016 in what can be described, at best, as egregiously fraudulent behavior and, at worst, as criminal behavior to boost immediate profits. That these highly regarded companies would engage in worse than what we have called "schlock" strategies should strike a cautionary note. The ease with which a company can slip into a schlock mode or worse is alarming.

In all three cases—Valeant, Wells Fargo, and Volkswagen—the remedy of removing the CEO was necessary but just a first step. Removal of the CEO in the case of Valeant was seen as a necessary gesture to appease public outrage but was not seen as necessarily improving Valeant's future behavior. The problem arose from Valeant's core beliefs and values. Valeant appeared to many to have become an irredeemable company.

In the case of Wells Fargo and Volkswagen, the removal of the CEO was seen as both a necessary and a promising first step toward redemption. The difference from Valeant was the assumption that the core beliefs and values of these companies were essentially sound, and that what was necessary was a CEO who embodied, embraced, and enabled the expression of these values.

The importance of the company's culture in transforming values into behavior cannot be overstated, and the role of the CEO in affecting the culture is of inestimable importance. The powerful roles played by Jack Welch at GE, Don Beall at Rockwell, Sam Palmisano at IBM, Steve Jobs at Apple, Jeffrey Brotman at Costco, and Howard Schultz at Starbucks are part of business legend.

---

* http://www.forbes.com/sites/kennethdavis/2016/06/03/a-market-fix-for-generic-drug-price-gouging/#3bc0c7487731

In today's world, the values that are gaining increasing acceptance and approbation are those described by Shalom Schwartz (2012) as self-transcendent, universal, and benevolent—in other words, humane, BoH values. It is apparent, therefore, that an alternative to schlock strategies and the temptation to engage in unscrupulous behavior in search of immediate profit gains is to take up the challenge of developing BoH Strategies. The BoH approach, though it demands more of managers and the leadership of companies, offers a wholesome, inherently satisfying, societally supported, and powerful approach to achieving both short-term and long-term economic and social sustainability.

The point to remember, however, is that for BoH Strategies to be adopted and effectively implemented, the CEO has to be a passionate champion of humanity—humane values and humankind.

THE FOUNDATION OF BoH STRATEGIES

The courage and commitment required to undertake BoH Strategies comes from either a value system that embraces social responsibility or a belief system that caring for people and the planet supports economic sustainability. Two companies, Dow Chemicals and Arvind, powerfully illustrate the needed value system and belief system.

Some aspects of Arvind have been discussed in earlier chapters. Though Arvind is a public company, the top management and leadership of the company has remained with the founding family through four generations. The current CEO, Sanjay Lalbhai, spoke* to us about the evolution of his family's enduring recognition of and commitment to social responsibility. It started with giving to good causes and then setting up trusts (i.e., charitable foundations) to ensure that monies were effectively employed, and finally incorporating the current approach, which is to integrate improved social well-being into Arvind's business models. Lalbhai is representative of corporate leaders whose personal value system leads to a commitment to BoH Strategies.

Dow Chemicals' commitment to BoH Strategies derives from a somewhat different premise. As articulated by Pedro Suarez, president of

---

* See: https://www.youtube.com/watch?v=VPC8Kn_XaSA

Dow Latin America, Dow's "purpose is human" and its means of achieving its purpose is its competency in science, specifically chemical engineering. One of Dow's fundamental values is a commitment to "protect the planet." In a fascinating presentation,* Suarez describes Dow's BHAGs as sustainability related and convincingly explains his firm belief that "sustainability is strategy." An interesting and related fact is that Suarez's territory was the most profitable in the company.

THE INFRASTRUCTURE OF BoH STRATEGIES

BoH Strategies demand the conscious integration of social benefit into the organization's business model. The integration of social benefit into the business model intrinsically requires an innovative approach. Innovative business models can be developed in five ways.

The first approach to promote innovation is to adopt BHAGs or stretch goals that include socially beneficial outcomes in addition to increased profits. This approach tends to lay the burden of creating innovative models on levels of management below the top management team.

The second approach is for the top management team to nurture or embrace disruptive technologies. A truly disruptive technology would make existing business models in the industry obsolete. An approach to identifying or developing disruptive technologies, which is advocated by Hart and Christensen (2002), is to focus on meeting the needs of the base of the pyramid, especially in emerging economies. Hart and Christensen's prescription is particularly promising and appropriate because addressing such needs is inherently socially responsible.

The third approach is a corollary of creating disruptive technologies by focusing on the bottom of the pyramid in emerging economies. Doing so, as we discussed in an earlier chapter, promotes reverse innovation (Govindarajan and Trimble 2012; Immelt et al. 2009)—employing the disruptive technology to gain competitive advantage in developed economies. Of course, the possibility of glocalization exists—modifying technology and products from developed economies to serve the needs of

---

* See: https://www.youtube.com/watch?v=hvuJp1hyD3Y&feature=youtu.be

emerging economies. Essentially, this can be generalized to the practice of connecting innovation ecosystems in the same industry across developed and emerging economies.

The fourth approach is to promote and take advantage of the explosion of innovation that results from connecting two industries. This phenomenon is strikingly evident today in the interaction of industries such as information technology and automobiles, and information technology and health.

The fifth approach is to tap into the creativity and different perspectives of stakeholders across the entire value chain. Engaging diverse stakeholders whose priorities and values might differ can be supported by the organization's avowed intent to share economic value added fairly with the stakeholders.

THE ESSENCE OF BoH STRATEGIES

Based on the preceding discussion, it follows that the fundamental core, the quintessence of BoH Strategies, would typically be characterized by the following:

1. An organizational **values and belief system** that is committed to and is motivated by the **symbiotic relationship** between **economic value added** and **social benefit**
2. The creation of **innovative business models** by
   a. Adopting **BHAGs** ("Big Hairy Audacious Goals")
   b. Employing **disruptive technologies** that offer the potential to address basic human needs at the base of the pyramid
   c. An empathetic response to **human needs**, especially at the **base of the pyramid in emerging economies**
   d. Connecting **innovation ecosystems** across
      i. Different **industries** to spur innovation
      ii. Developed and emerging **markets** to stimulate and catalyze reverse innovation
   e. Engaging stakeholders across the value chain—with a special focus on customers/clients—in **co-creating value**
      i. A related commitment to **sharing economic value** added with partners and stakeholders across the value chain

DEMONSTRATING THE DEVELOPMENT
AND IMPLEMENTATION OF A BoH STRATEGY

As part of the BoH Project,* we have been engaged in developing and implementing a BoH Strategy. This initiative has been labeled DC-HEaRT (Direct Current for Humanity, Energy and Regional Transformation).† Our approach in designing and implementing the DC-HEaRT Initiative maps onto and illustrates the quintessential characteristics of BoH Strategies listed immediately above.

We will describe the DC-HEaRT Initiative of the BoH project and then examine how it maps onto the fundamental characteristics of BoH Strategies. The DC-HEaRT Initiative had its origins in the third international BoH conference that was held in Pittsburgh in 2011. The primary purpose of the meeting was to review the progress that had been made in developing the theoretical underpinnings of the BoH Proposition, especially through the study of carefully selected organizations in the United States, Brazil, Russia, the Czech Republic, and India. There was a general consensus that empirical evidence and existing theories provided ample support for the BoH Proposition. At that meeting, the challenge was posed and accepted to design and invest in a project that would serve as an illustration of the process of developing BoH Strategies and also serve as a further demonstration of the power of the BoH Proposition.

Two of the co-authors of this book, who were the founders of the BoH Project and who continue to serve as the principal investigators, undertook this task. We will now describe how we crafted the DC-HEaRT Initiative to map onto and demonstrate each of the fundamental characteristics of a BoH Strategy.

Regarding **Characteristic 1** of a BoH Strategy, it is apodictic that as the principals of the BoH project, we met the first requirement—the quintessential characteristic of "*an organizational **values and belief system** that is committed to and is motivated by the **symbiotic relationship** between economic value added and social benefit.*"

In line with **Characteristic 2(a)** ("*adopting BHAGs*"), based on extensive discussions with colleagues, we chose the BHAG of "regional (economic) transformation." Our aspiration was to enhance the quality of life of selected communities by accelerating their economic development.

---

* See: www.katz.pitt.edu/boh
† See: http://www.dcpower.pitt.edu/

Given our physical location, existing partners and contacts, we decided to address underserved communities in the city of Pittsburgh in the United States and in the state of Gujarat in India. Our decision to choose these two communities was also motivated by the desired characteristic 2(d) ii—connecting "developed and emerging **markets** to stimulate and catalyze reverse innovation."

The next step, to meet *Characteristic 2 (b)*, was to identify a "*disruptive technology that offers the potential to address basic human needs at the base of the pyramid.*" One of the early possibilities that we considered was identified by Professor Ashok Jhunjhunwala of the Indian Institute of Technology, Madras, a long-standing partner in the BoH Project. The Indian Institute of Technology, Madras (IIT-M), is one of India's elite engineering schools, analogous to MIT and Caltech in the United States. Dr. Jhunjhunwala's renowned and successful efforts to bring communications-technology-based benefits to rural India are documented in a video* available on the BoH website.

At the aforementioned third conference of the BoH Project, Dr. Jhunjhunwala talked about his shift away from bringing communication to rural areas in India. In light of the explosion of cell phone usage and low-cost network availability in India, he concluded that his communications goals for rural India had been met.

Dr. Jhunjhunwala's BHAG now was (1) to combat the endemic brownouts and blackouts that plagued Indian cities and (2) to bring electricity to the thousands of off-grid or communities in India. He had carefully identified direct current (DC) power as the technology that would enable him to meet both goals. He had chosen to provide DC power because of the many advantages that DC offered over alternating current (AC). AC has been the global standard ever since the large-scale generation of electric power started in the early twentieth century. Widespread adoption of DC would be a significant disruption to the dominant technology.

Two years later, in the fourth BoH conference on "Realizing the Promise of DC Technology: 'Energizing' Low-Income Communities," held in Prague in October 2013, he presented the innovative approach that he proposed for combating the daily hours-long brownouts and blackouts in most second-level cities and towns on the AC grid. His team of faculty and students at IIT-M had designed controllers that could sense imminent

---

* http://www.katz.pitt.edu/boh/case-studies/bttm-pyramid.php

brownout or blackouts and shift the supply to each affected home to an unfailing 100 watts of DC power—enough to power several, highly efficient LED lights and a DC-powered fan. The reduced load on the grid would preempt the brownout or blackout and provide each home with essential power. Within a year after he presented his approach at the BoH conference, the Indian government and four state utilities invested in large-scale rollout of his technology.

To achieve his second goal of providing power to off-grid communities, he again proposed to resort to DC power generation. He planned to use solar panels, which generate DC power and allow for low-cost power generation anywhere there is sunlight. Moreover, LEDs, which are DC powered, are vastly more efficient than traditional incandescent lighting. Furthermore, efficient DC motors have been developed to power fans that were desperately needed in homes to combat the tropical heat and the indoor pollution from cooking with firewood and dried cow dung.

While DC power may be viewed as a natural fit with the existing context in rural India, in the United States, it would clearly function as a disruptive technology, because of the existing, universal availability of AC power on the national grid. During our initial discussions regarding the potential of DC power to serve as the basis for innovative business models intended to serve basic human needs in the United States, the idea was met with varying degrees of skepticism. Clearly, DC power met basic human needs in rural India, but there was no reason to anticipate that it could provide an economically justifiable basis to supplant AC power in meeting basic human needs.

The national power grid in the United States employs AC. Nikolai Tesla and George Westinghouse, the proponents of AC power, won the argument with Thomas Edison, the champion of DC power, because of the ease with which AC could be transformed into very high voltages to reduce transmission losses over long distances from mega power plants to consumers in distant cities.

However, we somewhat serendipitously discovered that the reasons why AC power was adopted in the United States were being chipped away. This initial advantage of AC power in the United States was being eroded for a variety of reasons:

- The mega plants generating AC mostly used fossil fuels, which are seen as a major contributor to global warming and potentially catastrophic climate change.

- The national AC grid was becoming increasingly unreliable because of growing power demands and chain-reaction type of failures.
- The aging AC grid is also seen as increasingly vulnerable to natural disasters and acts of terror.
- With the advent of the power transistor and breakthroughs in circuit breaker design, it is now possible to efficiently transmit DC power over long distances. Furthermore, such DC transmission is more efficient than AC because it can be accomplished with two wires instead of three and the diameter of these wires could be reduced because the skin effect (power being transmitted in the outer circumference of the wire) caused by AC does not happen with DC power.
- Sources of DC power such as photovoltaic panels are becoming more efficient and less expensive.
- Photovoltaic panels and other renewable sources of DC power can be sited in locations where the power is needed and also largely eliminate the concerns about global warming, which is exacerbated by large-scale power plants running on fossil fuels.
- Locally generated power reduces the pressure of growing demand on the aging AC grid. It increases the reliability and resilience of the national electric power supply.
- All electronic equipment requires DC power. The explosive increase in the use of electronic equipment—TVs, computers, smart phones, the Internet of things—results in a proportionate increase in losses resulting from the need to convert AC power from the grid to DC. These losses, at the low end, are in the range of 11% to 15% of the equipment's power requirement.
- Government subsidies and incentives to support renewable energy make the economics of DC power more attractive.

As we explored the potential of DC power as a disruptive technology in the United States, it became increasingly clear that it is likely that DC had the potential to gradually supplant AC in the long run. Our focus, however, was in the immediate and short term. Our concerns about committing to DC as a disruptive technology were drastically reduced when we, again serendipitously, discovered that our partner in Europe, the EU-supported Nupharo innovation park, was planning to focus on developing DC power applications, largely focused on middle-class needs in Europe.

A third development led us to commit to DC power as the disruptive technology of choice for our demonstration project. One of our clients,

Universal Electric Corporation (UEC), made a strategic decision to develop a business unit to engage in the market for systems of locally generated DC power that are now being designed and implemented by the largest consumers of electric power in the world—computer data centers.

Essentially, over a period of months, we had "triangulated" on DC power as the disruptive technology on which we would base our business models. IIT-M's use of DC power to address basic human needs at the base of the pyramid in India, Nupharo's development of DC power applications for the middle class in Europe, and UEC's investment in DC power to meet apex needs informed our choice of DC power as a potentially "powerful" stimulus for BoH Strategies.

Characteristic 2(c), *"an empathetic response to **human needs**, especially at the **base of the pyramid in emerging economies**,"* was supported by selecting the most underserved communities in Gujarat state in India, and in the city of Pittsburgh in the United States. While both Gujarat state and the city of Pittsburgh have reputations for being the most advanced, forward-looking, and well-managed political entities in India and the United States, respectively, our research and our on-site explorations enabled us to locate deeply disadvantaged communities in both locations. In Gujarat state, our local partner, the Narottam Lalbhai Rural Development Fund (NLRDF), identified seven villages, with more than 250 households, which have no access to power despite Gujarat state being one of the most advanced in India in rural electrification. After conducting a census and based on discussions with NLRDF, we chose to focus initially on 50 households in the village of Tuvar, in the Khed Brahma district of Gujarat state.

Similarly, our partner and colleague in the School of Social Work at the University of Pittsburgh—Dr. John Wallace, Professor of Social Work as well as of Business Administration and of Sociology—identified the Homewood neighborhood as a disturbingly disadvantaged community in Pittsburgh and connected us with a significant faith-based organization located there.

Empathetically identifying the basic needs of the community in Tuvar initially appeared to be obvious—they needed electric power to bring them literally and metaphorically out of the dark ages into, perhaps, the early part of the twentieth century. The census information, visits to Tuvar, and meetings with the villagers organized by NLRDF, however, led to a more insightful appreciation of the community's complex needs. The family income of many of the villagers was as low as US$0.50 per day. Illiteracy and health problems were found to be rampant. The nearest clinic was miles away over largely dirt roads. Many of the villagers, particularly the women, had never visited a large city or even a town. Premature aging caused by malnutrition and chronic diseases was accepted as the natural condition. The source of water for all personal needs was a single hand pump located a couple of hundred yards away downhill from the nearest abode. No sanitary facilities existed, either indoor or outdoor. The one- and two-room homes were dark even at midday because of the lack of windows. In the weeks before one of our visits, a young boy had succumbed to the bite of a poisonous snake, which he did not notice had entered the home, because it was so dark. The air within the homes was oppressive, poisoned by the cooking fires and the guttering flames of smoky lamps fed by adulterated kerosene. The only electrified building in the village was the tiny two-room school. None of the villagers had any idea whether the single, ancient computer in the school was connected to the Internet.

The most compelling needs that the visits and conversations revealed, which would otherwise have been outside of the awareness and comprehension of the BoH principals, stemmed from the distressing condition of women in the village. The lack of water and sanitation facilities resulted in particularly cruel and dangerous conditions for the women. The lack of economic opportunity in the village led some of the menfolk to seek employment in towns and cities, often resulting in them being away from home for extended periods. When the men returned, it was not unexpected for some of them to be carrying the AIDS virus into their homes.

Once an appreciation of the basic needs of the Tuvar community had been identified, the BoH principals engaged in an intense search for a partner who could help with the design and implementation of a plan to meet these needs. Potential partners were assessed on five criteria. First, were the values of these partners aligned with those of the BoH Project? Second, did they possess the complementary competencies needed to provide a systemic response, encompassing identification, design, and implementation, to the needs of the Tuvar community? Third, did they process the infrastructure to communicate effectively with multiple and diverse partners in the United States and India? Fourth, was the leadership personally compatible with the leadership of other partners, especially the BoH principals? Fifth, were they sensitive to and respectful of the cultural differences between the many partners likely to be involved?

With regard to the health needs of Tuvar, the choice was obvious and easy. The BoH principals had the advantage of connections and considerable familiarity with Apollo Hospitals,* the leading private hospital in India. With regard to DC power and the physical amenities needed in Tuvar, only one company was found that met all the criteria. During the course of months of interactions, this company—Safeworld Rural Services—proved to be an invaluable contributor not only in technical and design terms but also in helping the BoH principals to understand and respond better to the needs of the Tuvar community.

The response of the BoH Project and its DC-HEaRT Initiative to these basic needs using DC power was relatively easy to fashion. DC power generated by photovoltaic panels was the clear choice to provide energy for Tuvar. The plan that was developed was to provide LED lights, a DC-powered fan and an outlet for charging cell phones. Cell phones have become ubiquitous in India and have proven to be a powerful means of connecting people at the base of the pyramid to the nation's economy.

The choice was also made to have a central power-generating capability rather than provide individual households with solar panels as is the usual practice in government-subsidized programs in India. There were several reasons why this choice was made, but the primary one was to obtain efficiencies of scope and scale as several other community-based services where needed and planned for Tuvar. By providing an appropriately sized battery—with a charge controller stepping down the 380 V DC employed to minimize distribution losses to 12 V DC—the free rider phenomenon experienced when utilities are newly provided to a community was expected to be effectively controlled.

To provide clean and readily accessible potable water, a tube well was planned with a submersible DC pump to draw water from an aquifer adjacent to the village. A 10,000 liter storage tank was to be employed.

Ten centrally located toilets were planned for the 50 households and, especially to meet the hygiene needs of women, 10 private washing areas were added to the plan. Water for the sanitation and hygiene needs would be drawn from the 10,000 liter tank. Potable, filtered water would also be supplied at the central location.

The remote location of Tuvar village and lack of basic amenities meant that the health needs of the villages could not be met by setting up a clinic

---

* See: "Apollo Hospitals: Touching a Billon Lives" available at http://www.katz.pitt.edu/boh/case -studies/apollo.php

staffed with professional medical personnel. The decision was made to set up a custom-designed telemedicine facility, and in order to do so, the leadership of the Telemedicine Networking Foundation and the Tele Health Services of Apollo Hospitals was contacted. Apollo Hospitals is a pioneer and leader in telemedicine in India. Based on site visits and discussions with the founder and head of Apollo's telemedicine units, Dr. Krishna Ganapathy, and the vice president for program development, Prem Anand, a unique telemedicine facility was designed.

Paramedics drawn from the local population were to be trained by Apollo Hospitals to serve as the primary care responders and to operate the telemedicine equipment. A local hospital was assessed and selected to provide diagnostic care and prescribe treatments for patients needing care beyond what the paramedics could be expected to provide. Apollo Hospitals' network of doctors, located about 1000 miles away, was contracted to provide specialist care if needed. Response times of 8 to 10 min were guaranteed by the network. The equipment would be maintained by Apollo Hospitals with a guaranteed maximum downtime.

Based on an evaluation of the needs of the Tuvar population, appropriate equipment with a strong emphasis on sophisticated diagnostic capability but with ease of operation was specified. The facility was to be erected on a turnkey basis by Apollo Hospitals. Apollo was also to design educational programs on health to be provided in the local language for the residents of Tuvar. These were to be designed to enable the men and women in Tuvar to learn about healthy behaviors and to be able to distinguish between disease-caused symptoms and the frailties caused by aging.

Based on an appreciation of the situation in Tuvar, the BoH principals pressed for an emphasis on wellness, going beyond the traditional perspective on healthcare in rural India as limited to treatment for pain, fevers, and infections. The intent was to leapfrog the mindset on health to that of the developed world, where prevention and wellness are emphasized. This focus aligned with the thinking of Apollo Hospitals' leadership that the future of economically sustainable healthcare lay in wellness programs.

Apollo Hospitals viewed the Tuvar project as a major learning opportunity that could be efficiently replicated elsewhere. Apollo contributed the initial design and program management as well as the cost of training personnel and developing educational programs. The BoH Project contributed the capital equipment and NLRDF picked up the operating costs of the facility.

Though an empathetic response to the basic needs of the Tuvar community was believed, by the BoH principals, to have been captured by the planned portfolio of products and services, the business model was not yet complete because the project had to be made economically self-sustaining. The Tuvar community, Safeworld, and Apollo Hospitals provided the BoH principals with answers to the challenge of economic sustainability. Safeworld's competencies extended to agri-business and they developed a business plan that responded to the extremely restrictive controls that the Indian government placed on the purchase and use of agricultural land. The plan, in its essence, involved leasing land from local farmers that was not being cultivated because of lack of irrigation or because the land had been poisoned by the excessive use of fertilizer—both common problems in India. Safeworld's intellectual property included the processes by which poisoned land could be remediated within a few months at a very low cost. Also, they proposed the use of DC power to draw groundwater to scientifically and efficiently irrigate unused land, without further depressing the levels in the aquifers. Premium crops and produce were identified. The supply chain from seeds to market was designed. The business plan projected profitability that would significantly increase the income of the local community. It was expected that the earnings from a plot of five acres would provide eight individuals with a lower middle-class per capita income. Moreover, investors in the agribusiness were projected to receive a return on investment ranging from a low of 9% to a more likely 15%.

Apollo Hospitals contributed another significant approach to increasing the income to the community. A "Common Services Center" was proposed, which would connect the village community to government welfare programs and subsidies, as well as create government-sponsored, zero-balance bank accounts for individuals and families. The bank accounts would enable the villages to receive government welfare payments and would also enable NLRDF, Safeworld, and Apollo Hospitals to pay the fees and wages earned by the community much more easily. The Center would also provide the sophisticated ID card that the government required of individuals for them to access government programs. The operator of the Common Services Center, to be selected and trained by Apollo, would earn a living from the fees received.

The income generated by the agribusiness would enable the residents of Tuvar to pay for the services and products they receive, promoting a sense of dignity and reducing the free-rider phenomenon.

Empathetically responding to the needs of the Homewood community in Pittsburgh was accomplished by discussions involving, among others, representatives of the Homewood community; Dr. John Wallace of the School of Social Work and pastor of the Bible Center Church located in Homewood; Kannu Sahni, the head of community relations of the University of Pittsburgh; Steve Ross, the chief technology officer of Universal Electric Corporation; Scott Izzo, the director of the Richard King Mellon Foundation; Richard Piacentini, the executive director of Phipps Conservatory; Kristy Bronder, the program director of the BoH Project; and Dr. Bopaya Bidanda and Dr. John Camillus, the co-principal investigators of the BoH Project. Dr. John Wallace's intimate understanding of and detailed demographic data regarding the Homewood community guided the focus of the group to two fundamental challenges faced by the Homewood community.

First, Homewood met the classic definition of a food desert. Residents had no grocery store, no restaurants serving balanced meals, and no easy transportation for shopping and dining outside the community. Second, while subsidized housing was being made available through government programs, the cost of utilities was often too high for individuals to take advantage of the housing that was available.

The choice of DC power as the disruptive technology to be employed as the basis for innovative business models led the group to two key decisions. First, the group decided to design and build DC-powered greenhouses that would grow nutritious produce to be supplied to new local restaurants. The greenhouses were to be entirely off-grid for both electric power and water. Second, entirely off-grid DC circuits powered by solar panels would be designed and installed in a carefully selected sequence—homes, offices, and institutional buildings to create an independent DC power grid that would significantly reduce the cost of utilities by a combination of reducing the cost per kilowatt hour of power and reducing the consumption of power through the use of DC appliances.

A consortium of architects and contractors with expertise and experience in the design and construction of green buildings was created for the greenhouse. The capability to design the off-grid DC system did not exist in the business community in Pittsburgh. The system was designed by a newly formed company—Solar Cell—created by four students from the University of Pittsburgh, led by a graduate of the engineering school's doctoral program in electrical engineering, who had focused on DC technology. Solar Cell's leadership included a student, who had taken the course on the BoH taught at the business school, who connected the company to the Homewood project.

The greenhouse that was designed and built had many unique and innovative characteristics. It housed both hydroponics and aquaponics units. The Bible Center Church (BCC), to which ownership of the greenhouse was transferred, managed the agri-business based in the greenhouse. BCC also built a contemporary, attractive, destination coffee shop, and planned a restaurant to utilize the produce and the fish grown in the greenhouse.

The greenhouse became an iconic benchmark for such off-grid construction—with locally generated power and stored, harvested rainwater. It came to be viewed as having far more potential than just a greenhouse. In order to recognize its broader functionality, BCC and the BoH Principals decided to refer to it as a bio-shelter. The bio-shelter was to serve as the basis for subsequent variations in design, supported by frugal engineering processes. One such design developed by faculty and students from Robert Morris University, Carnegie Mellon University, and the University of Pittsburgh, was for tiny, off-grid, movable homes intended initially for veterans. The design was the basis for a business plan for building and selling affordable off-grid tiny homes that could be customized by the purchaser, thus promoting a sense of pride and ownership.

The DC circuits, controllers, and DC appliances needed to lower the cost of utilities were to be designed and provided by Dr. Jhunjhunwala and IIT-M. The experience and knowledge possessed by our Indian partners far outstripped what we were able to identify and access in the United States. Complementing the know-how of our Indian technologists, engineering students at the University of Pittsburgh designed and documented the processes for converting inexpensive AC appliances available in the United States to be run on DC power. Also, an Executive MBA team researched and identified alternatives to the appliances—especially in the heating, ventilation, and air conditioning arena—that could not readily be run on DC.

A critically important complementary project that was motivated by the BoH Project was the relocation of the Manufacturing Assistance Center (MAC) co-founded and directed by Dr. Bidanda, one of the principals of the BoH Project. Originally set up to train individuals with no technical skills for careers as machinists, the MAC was refocused to offer training programs leading to employment in DC-related construction, installation, and manufacturing industries. It also offered "maker spaces" and facilities for small-batch assembly lines.

Characteristic 2(d), *"connecting innovation ecosystems across: different industries to spur innovation, and developed and emerging markets to stimulate and catalyze reverse innovation,"* was designed into the DC-HEaRT Initiative from the very start. The two industries that were connected were energy and health. The two markets that were connected were the developed economy of the United States and the emerging economy of India.

Technology was to be transferred between the United States and Indian sites. Paradoxically, agriculture technology was to be transferred from the United States to India, and DC technology was to be transferred from India to the United States. Reverse innovation relating to DC technology and glocalization of management practices and agronomy were to take place.

The DC-HEaRT Initiative created innovation ecosystems in the United States and India including universities, management and engineering schools, vocational training organizations, small and large companies in the healthcare and electrical industries, philanthropic foundations and trusts, and professional associations in electrical engineering and health. The DC-HEaRT website (www.dcpower.pitt.edu) was the platform for a global community of knowledge and action focusing on DC power. Blogs and moderated forums were supported on this platform. The fifth BoH conference, focused on "Energizing Human and Economic Development: Leveraging the Power of DC Technology," was held in Pittsburgh in 2016. The conference brought this global community together to discuss, review, and guide the progress of the DC-HEaRT Initiative.

Characteristic 2(e), *"engaging stakeholders across the value chain—with special focus on customers/clients—in co-creating value and a related commitment to sharing economic value added with partners and stakeholders across the value chain,"* has been meticulously incorporated in the processes and outcomes of the DC-HEaRT Initiative. The client communities, the design and construction partners, the funding partners, and the knowledge partners were consciously and constantly engaged throughout the process. The economic value added was to be shared with stakeholders—communities, designers, contractors, operations managers, employees, funding agencies, and nongovernmental organizations (NGOs)—throughout the value chain.

The strategic underpinnings of the DC-HEaRT Initiative are illustrated in Figure 9.1.

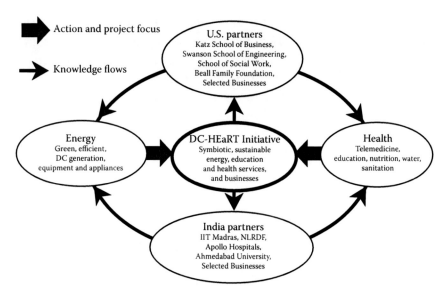

FIGURE 9.1
Strategic underpinnings of the DC-HEaRT Initiative.

LESSONS AFFIRMED AND LESSONS LEARNED

The DC-HEaRT Initiative of the BoH Project has proven to be an invaluable learning experience that has gone beyond our initial intent and hope that it would serve as an example of the process of developing BoH Strategies and would affirm the practicality and the value of the process. The practicality and value of the BoH formulation process, of course, in our estimation has been resoundingly affirmed in the course of designing and implementing the DC-HEaRT Initiative. Without our belief in the power of BoH Strategies and our understanding of the quintessential characteristics of such strategies, what we accomplished through our globally connected innovation ecosystems would not have been realized. In the process of designing and implementing the DC-HEaRT Initiative, our understanding of the nature of BoH Strategies was significantly enriched.

The Need for a Systemic Response

An important characteristic of BoH that we recognized but did not fully appreciate or stress adequately at the start of the DC-HEaRT Initiative is the implication of the inextricably tangled nature of the causes of the wicked problems that BoH Strategies seek to address. In attempting to respond empathetically to the community's needs, we came to understand and firmly believe that it is less than fully effective to tackle elements of a social problem in a linear fashion. This understanding was strongly reinforced by our experiences in Tuvar and in Homewood. Bringing electricity to homes in Tuvar undoubtedly improved the life and well-being of individuals. But the significance of providing lights and a fan pales somewhat when viewed in the context of the fundamental needs that are not met—potable water, sanitation, women's hygiene needs, healthcare, and the means to buy basic necessities such as food and clothing. Perhaps we can make this point by comparing what we sought to achieve in Tuvar with what Muhammad Yunus so brilliantly accomplished with his microfinancing programs. Yunus's programs, which fittingly earned the Nobel Peace Prize, focused on improving the economic well-being of an individual at a time. What we sought to achieve was to improve the economic well-being of an entire community, not just an individual, and furthermore to simultaneously provide all the members of the community with the means and opportunity to live a healthy, secure, societally connected, and more fulfilling life. Our time and interactions in Tuvar and our observations of the disappointing results of the well-intentioned but unidimensional efforts of many NGOs drove home to us the critical importance of providing a holistic response.

This need for a holistic response was also strikingly evident in Homewood. For instance, the inability of a majority of the children attending Homewood schools to take advantage of universally available scholarships to college provided by a program called the Pittsburgh Promise could be traced to a myriad of causes. For example, the fact that young boys and girls who were not served by school buses because they lived less than two miles from the schools, walked in the dark of the bitter early-morning winters in Pittsburgh, through a dangerous neighborhood, without having had breakfast, often tardy as a consequence, and therefore unable to meet the attendance requirements for receiving the benefits of the Pittsburgh Promise. In response, Professor John Wallace started a company that provided free transportation for these students. In true BoH fashion, he achieved economic sustainability by marketing transportation services to nearby institutions when his buses were not scheduled to drive students to and from school. After-school and meal programs were also organized by him.

The difference between Homewood and Tuvar was that in the former, multiple organizations, on their own initiative, provided a variety of services and the challenge was to effectively integrate the services provided. In the latter, the DC-HEaRT Initiative had to create the consortium necessary to address the entire set of basic needs of the community.

Our partners in Homewood, Professor John Wallace, and the BCC recognized the complexity of the challenge, and along with a consortium of foundations, companies and government agencies sought to provide the holistic response that the community required. The DC-HEaRT Initiative in Homewood could choose, with good reason, to focus on just the food desert characteristic of Homewood and the need for lowering the cost of utilities, because of the constellation of other services provided to the community.

In Tuvar—which had received little if any attention from government agencies and NGOs before its identification by NLRDF as the site for the DC-HEaRT Initiative—a very different situation existed. The BoH Project and NLRDF had to create a consortium—bringing in Apollo Hospitals, Safeworld, IIT-M, Ahmedabad University, local community leaders, and the nearby hospital—to provide the entire set of basic services that the community needed to grasp their potential as human beings in a civilized world.

Meeting the Challenge of Economic Sustainability

The lesson learned with regard to economic sustainability, which is at the core of the BoH Proposition, is that though it is initially a challenge, once an innovative business model has been devised and implemented, it can reasonably be expected to trigger a beneficial cycle of spin-offs and profitable growth.

First, let us consider why adopting a BoH Strategy is initially so challenging. While creating a BoH business model can be difficult, we learned that implementing an innovative strategy is just as challenging, perhaps even more so, than devising it. What we learned in the course of the DC-HEaRT Initiative reinforced what we knew and what every experienced manager knows—that it requires a high level of management expertise and commitment to create and implement an innovative business model. Managerial capability of a high order is essential to devise and implement BoH Strategies.

To devise a BoH Strategy, the ability to implement feed-forward processes is crucial. Building transformational and possibility scenarios, identifying robust actions, and adopting a real-options approach are necessary, advanced strategic management skills that are often not emphasized in business school curricula.

To implement a BoH Strategy, complementary levels of managerial skills are necessary. For instance, consider frugal innovation, which is critical to the effective implementation of BoH Strategies. The processes underlying frugal innovation—value engineering, quality engineering, and Lean engineering—require high-level professional expertise to implement effectively.

The reason why so many social entrepreneurship endeavors fail is because the need for management expertise is often overshadowed by the glow and justifiable satisfaction of embracing a socially beneficial value proposition. Indeed, in our experience, the multiple, sometimes intangible and complex objectives of social entrepreneurship demand an even higher level of managerial capability. Of course, this need for management expertise is just as pronounced in the context of BoH Strategies that seek to synergize economic and social goals.

In a conversation with the leadership of Fujitsu Corporation in Japan, several years ago, we learned two additional important lessons. First, the word "failure" is not part of Fujitsu's strategic management lexicon. Fujitsu understood that major projects will inevitably encounter unexpected challenges and setbacks and that managers' responsibility is to learn from and adapt to such challenges. Second, it became clear to us that there is as much a cultural underpinning for such a perspective as there is a management rationale. It was pointed out to us that, in some cultures, the United States being a prime example, the first instinct after a setback is encountered is to pin the responsibility on an individual or individuals and ensure some form of penalty or punishment. In fact, this cultural norm finds expression even in the dominant language of the culture. For instance, in the United States, it would be natural to say: "John did not meet the profit goals." In Japan, the expression is more likely to be: "the profit goals were not met." In the United States, therefore, the next step in the problem remediation process would be to determine the extent of blame and the appropriate punishment. In Japan, the next step would be to analyze why the profit goals were not met.

We hope that raising awareness of the cultural obstacles to implementing BoH Strategies will help reduce the magnitude of the challenge, and, of course, the frameworks, processes, and techniques that we have described should enhance management's capability to devise and implement BoH Strategies.

Having discussed the major sources of the initial difficulties in devising and implementing BoH Strategies, we can now examine why, after these initial challenges are met, BoH Strategies tend to accelerate the organization's economic sustainability progress.

Our experience with the DC-HEaRT Initiative in meeting the challenge of economic sustainability affirmed the importance of a pilot effort. The benefits of prototyping the business model cannot be overemphasized. We found, for instance, that while we sought to keep costs as low as possible, we could identify radical cost reduction possibilities only after implementing the first prototype. Our experience affirmed the lessons we learned from Fujitsu. It takes confidence and commitment to recognize that the typical oversights, cost overruns, and unexpected difficulties that go with building the prototype are inevitable and are the foundation for a further cycle of frugal engineering and strategic innovation.

The case of Arvind Ltd. is inspirational. Arvind's introduction of organic cotton farming to near-destitute farmers evolved naturally to the growing of food crops in periods when fields had to remain fallow, resulting over time in the development of a new and profitable agribusiness strategic business unit for Arvind. In the Homewood (United States) component of the DC-HEaRT Initiative, the prototype bio-shelter that was designed and built forms the technical foundation for a business engaged in the design of movable, off-grid tiny homes for veterans. In the Tuvar (India) component of the DC-HEaRT Initiative, working on the design of the telemedicine unit led to the identification of a Common Services Center, providing income for an entrepreneur and connection to government programs and banking services for the community.

Our experience suggests that the radical innovation that is intrinsic to BoH Strategies generates spin-off businesses. The more diverse the community and stakeholders connected to the projects, the more likely and varied the spin-offs that emerge.

THE INHERENT POWER OF BoH STRATEGIES

In Chapters 2, 3, and 4, we employed logic and analysis to develop

- An understanding of the critical challenges created by environmental mega-forces
- A recognition that effective responses to these critical challenges demand organizational strategies that recognize the needs of humankind and possess humane characteristics
- The design of a management framework that integrates these responses—the BoH framework that supports the innovative co-creation of an imagined future characterized by symbiotic economic and social sustainability

Logic and analysis are of vital importance. Logic and analysis serve to transmute disruption, conflict, and chaotic ambiguity into social and economic value added. Logic and analysis create the integrated BoH framework, which provides organizational inspiration, distinctiveness, and judgment, which incorporates planning processes that overcome complexity and uncertainty, and which provides a supportive foundation of frugally engineered value.

The strategies forged by the logically developed BoH Management Framework can certainly be expected to be powerful. But there is even more to BoH Strategies than the power and potential derived from the underlying rationality of the framework that creates them. We believe that BoH Strategies blend logic and analysis (the head) with human inspiration (the heart) to create a transcendental power.

Perhaps an example from another field might communicate what we mean more effectively. One of the leading groups of the 1960s was the Mamas and the Papas. Some of their iconic songs, such as *California Dreamin'*, *Monday Monday*, and Mama Cass's *Dream a Little Dream of Me* live on to today. One of the group members, Dennis Doherty, when reminiscing about their glory years, expressed what we are trying to communicate about the power of BoH Strategies. "Papa" Doherty claimed that the group of four, at their best, achieved a resonant harmony that created what he called a "fifth voice." The group actually gave this synergistic blending of their voices—this "fifth voice"—the name "Harvey!" "Harvey's" presence is especially evident in the group's rendering of *California Dreamin'*.

We are suggesting that BoH Strategies can also create "Harvey." Spotlighting, as we have, the three critical challenges created by the identified mega-forces enables managers to focus on the essentials, rather than being distracted by a plethora of issues. The BoH management framework is the crucible that transmutes these critical challenges to opportunities that beget the very real prospect of enhanced and symbiotic economic value and social benefit. BoH Strategies possess the potential for creating the "fifth voice," a resonant harmony resulting from the brilliant constellation formed of innovative business models, co-creating value and realizing imagined futures.

This "fifth voice" of the BoH is its heart—deriving from its core of recognizing and responding to humankind, and viewing strategic alternatives through the lens of humaneness.

The resonant harmony intrinsic to BoH Strategies is a natural result of the interdependent and mutually supportive nature of the key characteristics of the strategies. BoH Strategies, in essence, are expected to

1. *Empathetically address basic human needs.* Human needs are most pressing at the bottom of the pyramid, where the extremely low disposable incomes give rise to the second characteristic.

2. *Embrace a BHAG.* Because of the very low disposable income levels of the billions of human beings with pressing needs, economic value added and sustainability, which is also an expected and critically important outcome of BoH Strategies, become a daunting challenge. Economic sustainability becomes the BHAG that demands and inspires innovation and creativity when addressing the needs of human beings at the bottom of the pyramid.

3. *Innovate employing a disruptive technology.* To achieve the BHAG of economic sustainability, radical innovation in the business model is an imperative. Such innovation can be achieved by employing a disruptive technology supported by frugal engineering. Disruptive technologies, it has been argued, are best stimulated by addressing the needs of the bottom of the pyramid, which reinforces the first in this list of characteristics of BoH Strategies.

4. *Nurture connected innovation ecosystems.* The importance of innovation mandates the embrace of this approach to accelerating its happening. Connecting innovation ecosystems occurring in two different industries or in two different markets is known to stimulate innovation. While determining which two industries are the best

candidates for connection may be difficult, the markets in developed and emerging economies are obvious candidates. This again suggests that while addressing the needs of the base of the pyramid in emerging economies may be difficult, especially with the BHAG of economic sustainability, there are benefits to be obtained. An example of such a benefit is reverse innovation, which can improve the organization's competitive advantage in the developed market where such advantage may otherwise be difficult to gain.

5. *Integrate reverse innovation and glocalization.* Bringing disruptive, reverse innovation to developed markets to gain competitive advantage would be complemented by taking ongoing sustaining innovation relating to these new technologies back to the emerging market. An ongoing, bidirectional stimulation and transfer of innovation, supporting the development of new business models, would meet the twin BHAGs of enhanced and symbiotic economic value and social benefit.

The intimate interdependence of these essential characteristics of BoH Strategies integrates the "head" and "heart" and contributes to the resonant harmony that creates the "fifth voice."

THE BoH AND THE FUTURE OF MANAGEMENT

Framing and exploring the BoH Proposition has been a long and exciting journey for us. In the course of this journey, we have become increasingly convinced and are emboldened to assert that business strategies aligned with the BoH Proposition will become the norm. We believe that BoH Strategies represent the future of strategic management.

To support this bold and provocative assertion, we offer three videos. The videos we offer in support of our assertion are about Aravind Eye Hospital,* Arvind Ltd.,† and Dow Chemicals.‡ We employ the first two videos to initiate the discussions in our three-credit MBA course entitled: "The Business of Humanity®: Strategic Management in the Era of Globalization, Innovation and Shared Value." The third video serves as

---

* See: https://www.youtube.com/watch?v=65TLzg7GShw
† See: https://www.youtube.com/watch?v=VPC8Kn_XaSA
‡ See: https://www.youtube.com/watch?v=hvuJp1hyD3Y&feature=youtu.be

a capstone for the course. The course itself, in keeping with our thinking and belief, is billed as an elective in advanced strategic management. Over the last few years, we have offered the course with great success in the United States, India, and Brazil to MBA and Executive MBA students. The success of and enthusiastic reactions to the course reinforce our belief that the BoH Proposition has the potential to influence, fundamentally and positively, the practice of strategic management.

Going back to the three videos, Aravind Eye Hospital, which was briefly mentioned by us earlier, is described in colorful detail by Pavi Mehta, the grandniece of the founder of this famous success story. Aravind started with a unique "business model" that totally ignored profit making. Driven by spiritual tenets, the founder, Dr. Govindappa Venkataswamy, a physically disabled ophthalmologist, retired from low-paying government service in India, embraced a BHAG of eliminating blindness worldwide. He and other ophthalmologists in his immediate family started on this remarkable mission with a tiny, 11-bed hospital in a small town. Payment for the services of the Aravind Eye Hospital was entirely at the discretion of the patient. Indeed, Aravind actively sought out indigent patients from nearby villages who could not pay, and provided free transportation and accommodation in addition to diagnosis and treatment. This humane approach to meeting the needs of humankind across the world resulted, one could argue almost miraculously, in Aravind being identified as one of the 50 most innovative companies by the magazine *Fast Company*, and Dr. Venkataswamy becoming one of *Time* magazine's 100 most influential people in the world.

Economic success was achieved quite casually and incidentally, without formal attention to profit making, in the course of providing the social benefit that was the organization's *raison d'être*. Aravind's "business strategy" maps wonderfully well on to the characteristics of BoH Strategies. Aravind focused on meeting the needs of the base of the pyramid. Treating people unable to pay necessitated breakthroughs in cost reduction. Treating large volumes of people with the same medical procedures led to standardized processes, increased efficiency, and—which is typical of healthcare— resulted in improved outcomes and quality. Former patients became fervent promoters of Aravind and the well-to-do—who were willing to pay generously for additional comforts and services—flocked to Aravind because of their reputation for quality.

Aravind, in order to reduce costs, sought to bring the entire value chain in-house. Young women from local populations were recruited and trained as nurses in Aravind's unique processes. Medical residents from all over the globe were attracted to Aravind to learn from the best. Medical devices were manufactured in-house by Aravind. For instance, intraocular lenses needed for cataract surgeries, which were sold for hundreds of dollars in developed economies, were manufactured for a couple of dollars and exported to 110 countries!

The consequent economic success and growth that resulted from the embrace of basic tenets of a BoH Strategy, despite the lack of any conscious attempt to make profits, is convincing evidence of the power of the BoH Proposition.

We have discussed the company that is the focus of the second video—Arvind Ltd.—in earlier chapters. Arvind started out as typical for-profit venture in textile manufacturing. Viewing the video and listening to the CEO of Arvind, Sanjay Lalbhai, describe the evolution of his family's thinking about social responsibility is fascinating. We see a situation where, because of the values of top management, social benefit is gradually and profitably integrated into business models. This integration of social benefit gives rise to a sequence of innovative business models that enhance Arvind's economic value added. Arvind gets the same profitable result as in the case of Aravind Eye Hospital.

Do note that Arvind's case is quite different from that of Aravind Eye Hospital. Arvind was founded with a primary, if not sole focus on profit making. Because of the values of the founding family and their evolving understanding of the company's role in society, social benefit was thoughtfully and profitably introduced into the design of their business models, tapping profitably into the power of the BoH Proposition. In contrast, Aravind started with no interest in profit goals, but achieved great economic success because of the inherent power of BoH Strategies. In both cases, humanity-driven strategies resulted in their creating and tapping into the synergy of economic and social benefit.

In the third video, Pedro Suarez, president of Dow Latin America, employs logical analysis and experience to come to the conclusion and demonstrate that "sustainability is strategy." This understanding has led Dow, over time, to espouse "protecting the planet" as its *raison d'être* and commit to science as the means of achieving this strategic intent. The company's BHAGs are expressed in terms of sustainability, with the added expectation that profits be enhanced in the course of achieving these sustainability goals. Dow's strategy, especially in its Latin

American operations, is totally aligned with the BoH Proposition. To reinforce the point about the power of BoH Strategies, we note that Dow Latin America, under Suarez's leadership, was Dow Chemicals' most profitable territory.

Watching and hearing Suarez on the video, one can hear "Harvey"—the "fifth voice" created by the resonant harmony described by Dennis Doherty of the Mamas and the Papas. Suarez talks, for instance, about the joy that Dow's employees experience when they commit their personal time to enhancing environmental sustainability in the communities to which they belong. According to him, sustainability-related BHAGs are richer, more evocative, more motivating, and inspire creativity more powerfully than relatively nondescript and superficial profit goals.

Beyond the empirical evidence that we have encountered, which is presented in the videos and examples we have discussed, our belief in the power of the BoH derives from the observed evolution of value systems that was referred to in Chapter 3. We quoted Shalom Schwartz (2012) to the effect that the vast majority of cultures increasingly emphasize values characterized by benevolence and universalism, by transcendence and openness to change. It would follow that BoH Strategies, which are aligned with the direction of the worldwide evolution of values, will gain increasing acceptance and support. Aligning with these values and adopting BoH Strategies should get managers back to serving the "higher aims" described nostalgically by Rakesh Khurana who regretfully makes the case that "...the image of the ideal executive was transformed from one of a steady, reliable caretaker of the corporation and its many constituencies to that of the swashbuckling, iconoclastic champion of shareholder value...." (Khurana 2007).

Our conviction that BoH Strategies will become the universal norm is consistent with the belief that was originally articulated by Theodore Parker, a Unitarian Minister, and repeated by inspirational leaders such as Mahatma Gandhi, Martin Luther King, and Barack Obama—that "the moral arc of the universe, though long, bends towards justice." We, in our more limited sphere, have come to the passionate belief that in order to achieve economic sustainability, "the arc of profitable business bends towards humanity." It is our hope and expectation that the BoH Proposition, framework, strategies, and processes that we have offered will enable you to sail ahead of the curve. Bon voyage!

REFERENCES

Govindarajan, V., and Trimble, C. (2012). *Reverse Innovation: Create Far from Home, Win Everywhere*. Boston, MA: Harvard Business Review Press.

Hart, S.L. and Christensen, C.M. (2002). The great leap: Driving innovation from the base of the pyramid. *Sloan Management Review*, 44(1): 51–56.

Immelt, J., Govindarajan, V., and Trimble, C. (2009). How GE is disrupting itself. *Harvard Business Review*, 87(10): 56–65.

Khurana, R. (2007). *From Higher Aims to Hired Hands: The Social Transformation of American Business Schools and The Unfulfilled Promise of Management as a Profession*, Princeton, NJ: Princeton University Press.

Schwartz, S.H. (2012). An Overview of the Schwartz Theory of Basic Values. Online Readings in Psychology and Culture, 2(1). http://dx.doi.org/10.9707/2307-0919.1116

Index

managing and co-incentivizing
stakeholders, 83
market share, 81
mission statements, 78
recognizing and engaging diverse
stakeholders, 82–83
values, 79–80
iGate, 3
Indian Space Research Organization
(ISRO), 118
Innovation, *see also* Frugal engineering
and innovation
along entire value chain, 95–97
ecosystems, 60, 140
imperative of, 33
"reverse," 15
Institute of Electrical and Electronics
Engineers (IEEE), 61
Integrity, 11
ISO certification, 8
ISRO, *see* Indian Space Research
Organization (ISRO)

J

Jaipur Foot, 110
J.D. Power, 8
JVC, 31

K

Key result areas (KRAs), 4, 5

L

Launch window, 118
Lean engineering, 114, 119
Lilly, 3
Low-income markets, 15, 39, 53, 135
Low Voltage DC (LVDC) Forum, 61

M

Malcolm Baldrige award, 8
Management framework, 71–76
constructs, 72
diagram, 75
feed-forward processes, 73–74
frugal engineering, 74

identity, 73
integrating of constructs to create,
74–76
integrating of responses, 71–74
reverse direction, 75
Manufacturing Assistance Center (MAC),
151
Market share, 81
Marks & Spencer, 99
Matsushita, 31
McDonalds, 31
MedicalHome, 108
Mercedes Benz, 34
Mission statements, 78
Multinational corporations (MNCs), 16
Murugappa Group, 3

N

NASA, 118
NASSCOM, 3
Nokia, 29
Nongovernmental organizations (NGOs),
19, 83
North American Philips, 30

O

Off-grid construction, iconic benchmark
for, 151
Open sourcing, 56
Organic farming, 47

P

PEST (political, economic, social, and
technological) model, 36
Philips, 30
Pittsburgh Brewing, 59
Pittsburgh Promise, 154
PMT choices, *see* Product–market–
technology (PMT) choices
Porsche, 34
Power grid, DC, 150
Product–market–technology (PMT)
choices, 78
Proposition, 1–26
BoH as dominant strategic mindset,
23–25

About the Authors

Dr. John C. Camillus earned his Doctor of Business Administration degree at Harvard University; his MBA at the Indian Institute of Management, Ahmedabad (IIMA); and his B. Tech degree in mechanical engineering from the Indian Institute of Technology, Madras.

He has held the Donald R. Beall Endowed Chair in Strategic Management at the Katz Graduate School of Business at the University of Pittsburgh since 1991. Before joining the University of Pittsburgh, he was professor of Management at IIMA.

He has published extensively in management journals, including *Harvard Business Review*, *Academy of Management Review*, *Strategic Management Journal*, and *Management Science*. He has authored several books including, most recently, *Wicked Strategies: How Companies Conquer Complexity and Confound Competitors* (Rotman-University of Toronto Press, 2016).

He has served on several boards of directors/trustees. He was a consultant to the top management of more than 100 organizations on four continents, including Fortune 500 companies, professional firms, and nonprofit organizations.

John has received numerous awards for teaching including, most notably, the University of Pittsburgh's "Chancellor's Distinguished Teaching Award" and the "Best Teacher" award at IIMA. He has received the "Distinguished Alumnus Award" from IIMA, the University of Pittsburgh's "Chancellor's Distinguished Public Service Award," and the Katz School's "Inaugural Diversity and Global Leadership Award."

 **Dr. Bopaya Bidanda** is currently the Ernest E. Roth professor and chairman of the Department of Industrial Engineering at the University of Pittsburgh. His research focuses on Economic Development, Manufacturing Systems, Reverse Engineering, Product Development, and Project Management. He has published nine books and more than 100 papers in international journals and conference proceedings. Recent (edited) books include books published by Springer Inc. on Virtual Prototyping & Bio-manufacturing in Medical Applications and on Bio-materials and Prototyping Applications. He has given invited and keynote talks in Asia, South America, Africa, and Europe. He also helped initiate and institutionalize the Engineering Program on the Semester at Sea voyage in 2004.

He has previously served as the president of the Council of Industrial Engineering Academic Department Heads (CIEADH) and on two rotations on the Board of Trustees of the Institute of Industrial & Systems Engineers. He also serves on the International Advisory Boards of universities in India and South America.

Dr. Bidanda is a Fellow of the Institute of Industrial Engineers and currently serves as a commissioner with the Engineering Accreditation Commission of ABET. In 2004, he was appointed a Fulbright Senior Specialist by the J. William Fulbright Foreign Scholarship Board and the US Department of State. He received the 2012 John Imhoff Award for Global Excellence in Industrial Engineering given by the American Society for Engineering Education. He also received the International Federation of Engineering Education Societies 2012 Award for Global Excellence in Engineering Education in Buenos Aires and the 2013 Albert Holzman Distinguished Educator Award given by the Institute of Industrial Engineers. In recognition of his service to the engineering discipline, medical community, and the University of Pittsburgh, he was honored with the 2014 Chancellors Distinguished Public Service Award.

 **Dr. N. Chandra Mohan** earned his doctoral degree in Economics from Jawaharlal Nehru University in New Delhi. He is a well-known business and economics commentator who has worked with all the leading newspapers and magazines in India including *The Hindu, Economic Times, Business India, Times of India, Financial Express,* and *Hindustan Times*—in a senior editorial capacity. He has also written for many other Indian publications like *Outlook, Business Standard, Free Press Journal,* and *Daily News and Analysis,* including international ones like *Fortune, Economist, FDI magazine* (in-house journal of the *Financial Times*), *Harvard Asia Quarterly,* and *Inter Press Service.*

In 1990, he was appointed a Parvin Fellow at the Woodrow Wilson School at Princeton University. He has also taught economics and international business to MBA students at multiple universities in New Delhi. He was previously a consultant with think-tanks like the Institute for Studies in Industrial Development and Research and Information System for Developing Countries. He has contributed to the research output of these think-tanks in the form of working papers and discussion papers. He has also contributed to research on economic diplomacy on topics like India's FTA with ASEAN, blue economy, energy, and other initiatives with the Society of Indian Ocean Studies.